Carpentry and Joinery

2

Second Edition

Brian Porter LCG, FIOC
Department of Building Trades, Leeds College of Building

General Technical Adviser and Consultant:
W. R. Rose MCIOB, DMS, DASTE, FIOC

Woodworking Machines:
Consultant and Contributor: Alan Wilson LCG, MIM WOOD.T
Consultant: Eric Cannell

BUTTERWORTH
HEINEMANN

OXFORD AMSTERDAM BOSTON LONDON NEW YORK PARIS
SAN DIEGO SAN FRANCISCO SINGAPORE SYDNEY TOKYO

Butterworth-Heinemann
An imprint of Elsevier Science
Linacre House, Jordan Hill, Oxford OX2 8DP
225 Wildwood Avenue, Woburn MA 01801-2041

First published by Arnold 1984
Second edition by Arnold 1991
Reprinted by Butterworth-Heinemann 2001, 2002 (twice)

British Library Cataloguing in Publication Data

Porter, Brian *1938*
Carpentry and joinery.   2nd ed.
Vol. 2
1. Carpentry and Joinery.
I. Title
694

ISBN 0 340 50774 8

For more information on all Butterworth-Heinemann publications please
visit our website at www.bh.com

Typeset by
Rowland Phototypesetting Ltd, Bury St Edmunds, Suffolk
Printed and bound in Great Britain by The Bath Press, Bath

Carpentry and Joinery 2

Libraries and Information

Carpentry and Joinery 2

Contents

Foreword

Preface

Welcome to the revised second edition of Brian Porter's textbook *Carpentry and Joinery 2*. This is not a novel to be read in one or two sittings, but a text and reference book to be used in conjunction with academic and practical training. Sections of the book should be used and referred to as and when education and training needs dictate. Looking at the illustrations and reading the text alone will not make a carpenter and joiner, but used in conjunction with study at college and practical experience at work it will assist in the development of that knowledge which underpins and reinforces true craft skills.

One of the biggest changes in the latter half of the twentieth century has been the codifying of materials and craft practices into standard codes, and Brian Porter makes full reference to these in his book. The adoption of common European standards in 1992 and the eventual adoption of International Standards will no doubt be the impetus for a further revision.

Mr Porter is a practising lecturer with a good understanding of the technical and practical aspirations of carpentry and joinery craft students which he has tried to meet, and I think succeeded, in this revised book.

W. R. Rose MCIOB, DMS, DASTE, FIOC.

Preface to Second Edition

With the advent of the National Vocational Qualification (NVQ) certification scheme, where each topic or subject area forms the basis of a testing module to be taken as and when required by students to meet their individual needs, textbooks such as this one will be found invaluable – particularly if such a book is easy to follow, easily read with a high visual content, and provides adequate answers to readers' questions. I feel confident that this book will meet all these requirements.

As with the first edition, this book has been designed to complement *Carpentry and Joinery 1* (hereafter referred to as Book 1) and where necessary should be read alongside it. In this way full benefit can be gained from the relevant cross-references. However, for those students who have already acquired basic craft knowledge, the reference between Book 1 and Book 2 may not be required.

Brian Porter

Preface to First Edition

This volume is the second in a series of three for students of carpentry and joinery. Volume 1 dealt in the main with the City and Guilds of London Institute part-1 craft certificate course number 585; volumes 2 and 3 jointly were to have covered the craft theory and associated subjects set down in part 2. However, in 1983 CGLI published a revised scheme for the 585 course which does not divide the craft certificate work into part-1 and part-2 stages but, except for a few topic adjustments, retains the previous subject matter. Collectively, volumes 1 to 3 of this series are in line with these changes.

Many areas of work covered in this volume were introduced in volume 1 – here, these have been either extended or developed to a higher level of technology. Duplication between volumes has been avoided, but, where clarification of a particular point or situation is needed, cross-references have been provided within the text.

Traditionally, building crafts have been learnt on a time-serving basis – usually in the form of an apprenticeship to be worked over a set number of years. Some sectors of industry, including our own, now seem to favour a different approach to achieving craft status – that of allowing their trainees to prove themselves by participating in a structured programme of skill testing. Skill tests will be designed to enable students to progress according to their individual abilities. Textbooks of this nature should, therefore, become an even more valuable asset to the student – not only as reinforcement material for current course work, but also as reference for future years.

Throughout this volume I have endeavoured to incorporate new materials alongside modern techniques. However – as in all building crafts, but particularly the timber trades – quality in the work will not materialise without the individual skill of the craft worker. Perhaps it could be said that the good craft worker is one who signs his or her work with pride.

Brian Porter

Preface to First Edition

Acknowledgements

I would again like to thank colleagues and friends at the Leeds College of Building for their help and guidance in compiling the first edition – in particular, Mr E. Judkins, Mr H. K. Servant, and Mr R. A. Wheeler for proof-reading and Mr B. Crossfield and Mr P. S. Seed for general assistance. I am also grateful to James C. Coulson, the North Eastern Regional Officer of TRADA, for his comments on Chapter 1.

For this new edition, I am indebted to Mr W. R. Rose (Assistant Principal at Leeds College of Building) for currently undertaking the role of technical editor, and consultant advisor. I would like to thank Alan Wilson LCG, MIM WOOD.T, for his work in updating the text and artwork on woodworking machines, and taking the photographs for Figures 3.1 and 3.47(b) and drawing Figure 3.45, together with the help given by Eric Cannell who also kindly proof-read work relating to timber technology, Neil Townend for his help on 'timber preservation' and David Smith and library staff at LCB for their interest shown and help given.

I would also like to thank Paul Monro (LCB), Brian De Lucchi (Leeds City Council Highways Department), Mr J. Ware (Kango Ltd), Mr P. Chamberlin (Thomas Robinson Group plc), Mr G. Wilshaw (Hilti (Great Britain) Ltd), Mr D. A. Biddlecombe (GKN Kwikform Ltd), Mr J. Bonnet (Elu Power Tools – Black and Decker Professional Products division), Mr T. Waldram (Atlas Copco (Great Britain) Ltd) and Mr K. Davis of South Field College of Further Education for their help by contributing to and or checking text and artwork. I am grateful to the following organisations for supplying information and, where noted, their kind permission to reproduce photographs or illustrations:

Acrow plc (Fig. 10.12); Atlas Copco (Great Britain) Ltd (Figs 2.29 and 2.30); Bevplate; Catnic Components Ltd; The Chipboard Promotion Association Ltd; Fosroc Ltd; GKN Kwikform Ltd (Figs 11.4 and 11.17 to 11.19); Hall & Watts Ltd; Hilti (Great Britain) Ltd (Figs 2.35 and 2.48); The International Truss Plate Association (ITPA); Kango Ltd (Figs 2.1, 2.2, 2.7, 2.9, 2.11, 2.14, 2.15); Makita Electric (UK) Ltd (Figs 2.24 to 2.26); Miller Homes Yorkshire Ltd; Mowlem Northern Ltd (Figs 5.8, 5.11, 10.6, 10.10, 10.13, 10.14); Optical Measuring Instruments Cowley Ltd; Parkside Superstructures Ltd; Protim (Fig. 1.53); Protimeter Ltd (Fig. 1.16); Rabone Chesterman Ltd, a division of the Stanley Works (Fig. 4.27); Rapesco Group plc (Fig. 2.23); The Rawlplug Company Ltd; Rentokil Ltd (Figs 1.42 to 1.44 and 1.46 to 1.52); Stanley Tools Ltd (Fig. 9.7); Wadkin Group of Companies (Figs 3.31, 3.37, 3.43); G. F. Wells Ltd (Figs 1.22, 1.25 to 1.28).

The scanning electron micrographs in Fig. 1.3 are reproduced by kind permission of Stewart J. Kennmar-Gledhill of the Imperial College of Science and Technology, London, and those in Figs 1.4 to 1.6 by courtesy of David Kerr and Barrie Juniper of the Plant Science Dept., Oxford.

Table 1.6 is based on a Crown-copyright chart published by the Building Research Establishment and is taken from the TRADA wood information sheet *Moisture content in wood*, with the permission of both organisations.

The Woodworking Machines Regulations 1974 (Crown-copyright) are reproduced by permission of the Controller of Her Majesty's Stationery Office.

Figure 9.15 is a Crown-copyright extract from Building Research Establishment (UK) Defect Action Sheet DAS83 [excluding diagram (d)], and is reproduced by permission of the Controller, HMSO.

The following are reproduced from or based on British Standards by permission of the British Standards Institution, 2 Park Street, London W1A 2BS, from whom copies of complete standards may be obtained: Table 2.4 (BS 4078 : Part 2 : 1989), Table 10.1 (BS 8110 : part 1 : 1985), Table 11.4 (BS 5973 : 1981).

Finally, I must again thank my wife – Hilary Yvonne – for her continual help, patience, and understanding during the writing of this book.

Brian Porter

1

Timber and its associated problems

As part of their craft expertise, carpenters and joiners should be able to identify common commercially used timbers and manufactured boards to the extent that they also become aware of how these will respond to being (Fig. 1.1):

a) cut by hand and machine,
b) subjected to loads,
c) bent,
d) nailed and screwed into,
e) glued,
f) subjected to moisture,

g) attacked by fungi,
h) attacked by insects,
i) subjected to fire,
j) treated with preservatives, flame retardants, sealants, etc.,
k) in contact with metal.

By and large, behaviour under these conditions will depend on the structural properties of the timber and its working qualities, strength, resistance to fungal decay (durability) and insect attack, chemical make-up, and moisture content.

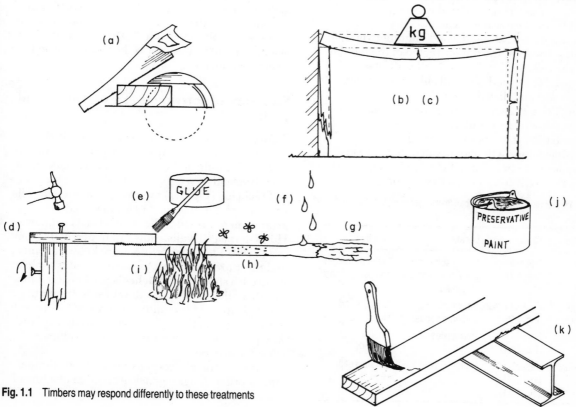

Fig. 1.1 Timbers may respond differently to these treatments

1.1 Timber identification and the structure of wood

Timber identification is not always as simple as it may at first appear. A generally accepted method of identification is to rely on the experienced craftsperson to compare colour, weight, smell, and texture with those of known species; however, because of the many different, yet similar species of wood used today, this method can result in guesswork.

Where true recognition is important – to comply with a specification or to ascertain how or why different species act or react under certain conditions – a more sophisticated method must be employed, involving the use of a hand lens (× 10 magnification) and/or a microscope of high magnification to look into how the wood is structured and what it contains.

Wood (*xylem*) is made up of successive bands of different forms of woody tissue (groups of similar cells with the same function – Table 1.1) distributed mainly along the axis of the tree's stem (axial cells) and to a lesser extent radially (radial cells) as shown in Fig. 1.2.

In many species – particularly those grown in a temperate climate – patterns of periodic growth known as growth rings (annual rings) are clearly defined by the formation of earlywood (springwood) cells and latewood (summerwood) cells. In softwoods, the darker region of the band is where the cells are smaller and cell walls have thickened. Hardwood cell walls likewise thicken, but their size and distribution also affect the pattern. Trees grown in tropical regions may not produce these rings annually, or may even not produce bands at all, because the climate responsible for growth may be such that continuous growth occurs, the only changes being those which reflect a wet or a dry season.

Table 1.1 Relationship between cell distribution, cell function, and wood type

Distribution	Tissue	Location	Function
Axial cells	Tracheids (Fig. 1.7)	Mainly softwoods, some hardwoods (e.g. oak)	Provide strength and conduct sap.
Axial and radial cells	Parenchymas (Fig. 1.9)	Softwoods and hardwoods	Conduct and store food.
Axial cells	Fibres (Fig. 1.11)	Hardwoods	Provide strength.
Axial cells	Vessels or pores (Fig. 1.11)	Hardwoods	Conduct sap.

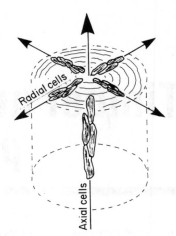

Fig. 1.2   Cell distribution

Ageing may produce another unmistakable feature – heartwood (duramen) and sapwood (alburnum). Heartwood is usually (but not always) darker in colour, due to chemical changes, inactive or dead tissue. It strengthens the tree by acting as its spine. The amount of heartwood will depend on the tree's age – the older the tree, the greater the amount. Sapwood (secondary xylem) is the active part of the tree, where cells conduct sap and store food. Young trees contain mostly sapwood tissue.

Figure 1.3 is the key which shows that from the smallest sample of wood it is possible – with the aid of a microscope – to see how the intricate wood structure is formed.

Structure of softwoods (Figs 1.4 and 1.7 to 1.10)

Softwoods belong to a group of plants called 'gymnosperms' (usually cone-bearing with naked seeds). The bulk of the 'stem' tissue is made up of axial tracheids – see Table 1.1.

Tracheids (Fig. 1.7) are elongated pointed cells, usually 2 to 5 mm long. Their interlaced arrangement provides the stem with its support. Holes in the cell walls allow sap to percolate from cell to cell. These perforations are known as 'pits', and there are two types: 'bordered' and 'simple'. Figure 1.8 shows how tracheids use bordered pits to either allow or stop the flow of sap. Injury or ageing (sapwood becoming heartwood) will result in the closure of these pits – which could explain why preservative penetration is often difficult within heartwood.

Softwood 'rays' – responsible for food storage, among other things – generally consist of a line of single radial 'parenchyma' cells (Fig. 1.9(a))

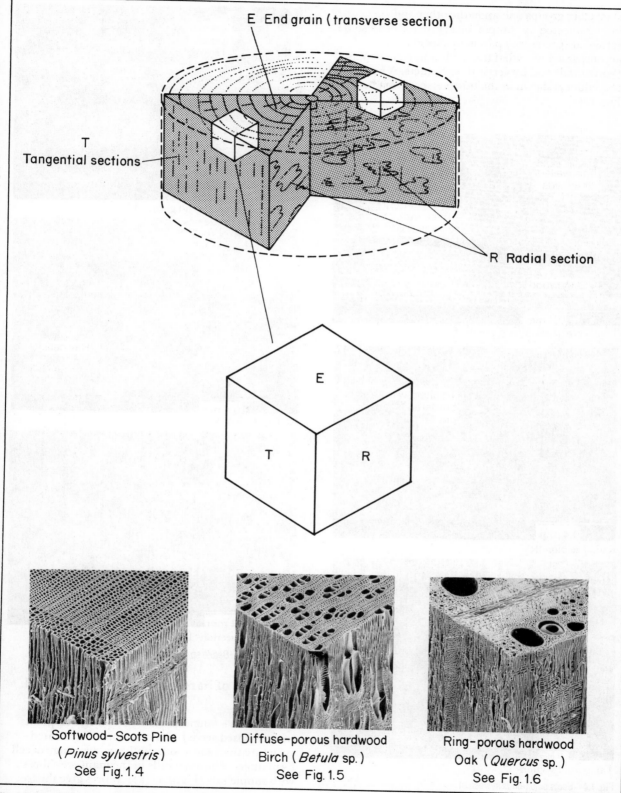

E End grain (transverse section)

T Tangential sections

R Radial section

E

T   R

Softwood– Scots Pine
(*Pinus sylvestris*)
See Fig. 1.4

Diffuse-porous hardwood
Birch (*Betula* sp.)
See Fig. 1.5

Ring-porous hardwood
Oak (*Quercus* sp.)
See Fig. 1.6

Fig. 1.3 Key to compare cell distribution (see Figs 1.4, 1.5 and 1.6)

stacked in groups one upon the other and interconnected by 'simple' pits (Fig. 1.10). In some softwood species, ray parenchymas may be accompanied by radial tracheids and resin canals (horizontal) and by vertical resin ducts. A vertical resin duct is shown in the transverse section in Fig. 1.4.

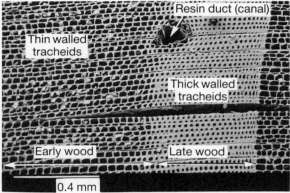

Transverse section (end grain) 'E'

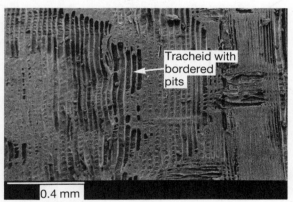

Radial section 'R'

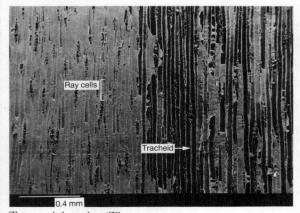

Tangential section 'T'

Fig. 1.4   Scots pine (*Pinus sylvestris*)

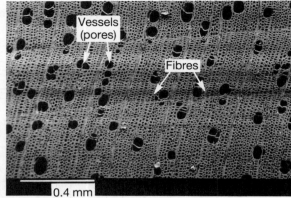

Transverse section (end grain) 'E'

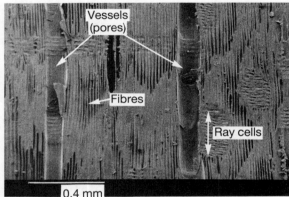

Radial section 'R'

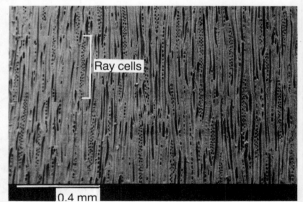

Tangential section 'T'

Fig. 1.5   Birch (*Betula* sp.)

Structure of hardwoods (Figs 1.5, 1.6, and 1.11)

Hardwoods – 'angiosperms' (broad-leaved plants with enclosed seeds) – have a more complicated structure than softwoods, with a wider range of cell formations. The axial cells in the main are 'fibres', with simple pits (Fig. 1.11(a)). Fibres give the tree

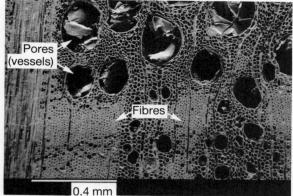

Transverse section (end grain) 'E'

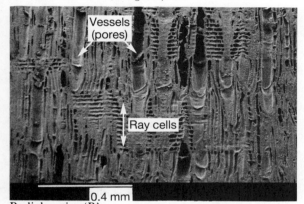

Radial section 'R'

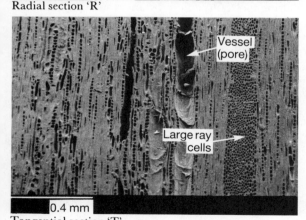

Tangential section 'T'

Fig. 1.6 Oak (*Quercus* sp.)

its strength. 'Vessels'.(Fig. 1.11(b)) – often also called 'pores' – are responsible for conducting sap and have both bordered and simple pits. As a means of distinguishing the difference between the terms 'vessel' and 'pore' the following explanation may be helpful:

The term 'vessel' may be used when the cell is cut through longitudinally and exposed across

either a tangential or radial section (Fig. 1.3). Conversely when it is exposed across a transverse section (end grain), (Fig. 1.3), it may be referred to as a 'pore'. In either case they are one and the same.

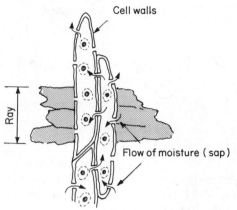

Vertical section

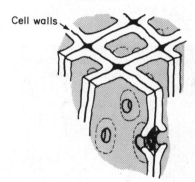

Perspective view

Fig. 1.7 Tracheids

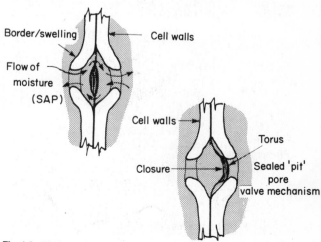

Fig. 1.8 Vertical section through a bordered pit
Closure due to stress – may be a result of ageing or injury

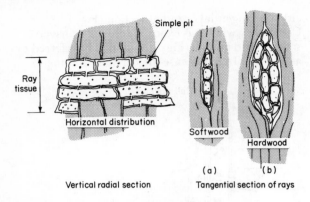

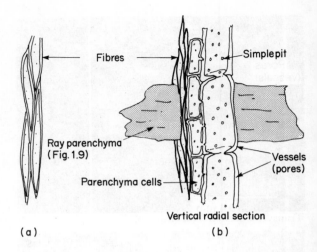

Fig. 1.9 Vertical section through parenchyma cells

Fig. 1.11 Hardwood cells

Rays in hardwood often form a very distinct feature of the wood. Unlike in softwood, ray tissue can be several cells wide as shown in Fig. 1.9(b).

Size and distribution of vessels are particularly noticeable in hardwoods with distinct growth rings. Two groups of these are as follows.

i) *Diffuse-porous hardwoods* (Fig. 1.5) These have vessels of a similar diameter and more or less evenly distributed around and across the growth-ring bands. Examples include beech, birch, sycamore, and most tropical hardwoods.

ii) *Ring-porous hardwoods* (Fig. 1.6) In these, vessels are large in the earlywood, then become smaller with the latewood. Examples include ash, elm, and oak.

Hardwood can therefore be classed as 'porous' wood. (Note: 'porous' is a botanical term only.) Not only do the pores of a hardwood provide an important distinguishing feature between the two types, the lack of them would indicate a softwood.

Cell contents

Once wood cells have been cut open, their exposed cavities may have undesirable contents in the form of calcium salts or silica grains, as in the case of some tropical hardwoods – for example, iroko (calcium carbonate) and keruing (silica). Silica grains can have a serious effect on the edge of cutting tools. Organic substances such as 'gums' in hardwood or 'resins' in softwood may also be present.

Other substances such as oils, acids, tannins, and latex are also to be found in wood. These are known as 'extractives'.

Woods such as oak, chestnut, Douglas fir, and Western red cedar contain acids which, under moist conditions, can corrode some metals. Woods containing tannins under similar conditions will react with ferrous (iron-containing) metals, the wood being stained a bluish-black colour.

Grain characteristics

'Grain' refers to the direction of the main elements of the wood. The manner in which grain appears will depend upon one or more of the following:

a) the direction of the cut,
b) the location of the cut,
c) the condition of the wood,
d) the arrangement of the wood cells.

Table 1.2 lists some common grain terms, together with a broad explanation of why they are so named.

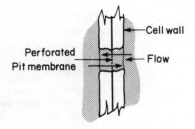

.N.B. <u>No</u> border or swelling

Fig. 1.10 Vertical section through a simple pit

Table 1.2 Grain terms and conditions

Grain terms and conditions	Explanation	Texture	Example of species	Remarks
End grain	Cross-cut exposure of axial and radial cell tissue	—	All	See Fig. 1.3.
Straight grain	Grain which generally follows a longitudinal axial course	—	Keruing Kapur	Hardwood and softwood
Cross grain	Grain which deviates considerably from being parallel to the edge of the timber	—	Elm	Hardwood and softwood
Open or coarse grain	Exposed large vessels, wide rays, and very wide growth rings	Coarse	Oak Ash	Associated with texture
Close or fine grain	Exposed small vessels, narrow rays and/or narrow growth rings	Fine	Sycamore softwood	Associated with texture
Even grain	Generally uniform, with little or no contrast between earlywood and latewood	Even	Spruce	Associated with texture
Uneven grain	Grain elements vary in size and uniformity – distinct contrast in growth zone	Uneven	Douglas fir	Associated with texture
Curly or wavy grain	Direction of grain constantly changing	Uneven	Walnut	Rippled effect
Interlocking grain	Successive growth layers of grain inclined to grow in opposite directions		African mahogany Afrormosia Sapele	Striped or ribboned figure
Spiral grain	Grain follows a spiral direction around the stem from roots to crown throughout its growth			Defect in timber, affecting structural use
Sloping grain (diagonal grain)	A conversion defect resulting from straight-grained wood being cut across its natural axial growth pattern, or a growth defect resulting from an abnormality in an otherwise straight tree			Defect in timber, affecting structural use (Book 1, page 23)

Note In addition to the above, short grain may result from timber being cut and may easily split due to the short length of its elements (e.g. within a trench sawn for a housing joint).

Texture

Texture is a surface condition resulting from the size and distribution of wood cells. Texture is usually associated with touch, but, unless the grain is filled, many surface finishes, especially high-gloss polishes, will reveal the texture – direction and intensity of light being the all-important factors. Texture is directly related to grain condition, as shown in Table 1.2, and is a typical characteristic of timber (Table 1.3).

Figure

Figure is best described as the pattern or marking which is formed on the surface of processed timber as a result of wood tissue being cut through. For example, quarter-sawn oak exposes broad rays, producing what is known as 'silver figure' (see Book 1, Fig. 1.18); the interlocking grain of African mahogany will reveal a 'stripe' or 'ribbon figure'. Tangential-sawn softwood, like Douglas fir, can show a very distinctive 'flame like' figure (Book 1, Fig. 1.18).

Timber possessing these characteristics can be regarded as having natural decorative properties – see Table 1.3, column (f), 'General usage'.

Table 1.3(a) Some properties of various softwood species

Colour of heartwood	Species		Origin	(a) Moisture movement	(b) Approx. density (kg/m³)	(c) Texture (see Table 1.2)	(d) Working qualities (cutting, nailing, etc.)	(e) Durability	(f) General usage
	Common name	Latin (botanical) name							
Pinkish brown	Douglas fir (Columbian pine)	*Pseudotsuga menziesii*	Canada USA & UK	Small	530	Fine/medium (straight grain)	Good	Moderate	Int. and ext. joinery*†
Pinkish brown	European Redwood (Scots pine)	*Pinus sylvestris*	Europe	Medium	510	Fine/medium	Good	Non-durable	Int. and ext. joinery*
Brown to red-dark streaks	Parana pine	*Araucaria angustifolia*	Brazil S. America	Medium	550	Fine (mainly straight-grained)	Medium to good	Non-durable	Int. use only Plywood
Light brown	Western hemlock	*Tsuga heterophylla*	Canada USA	Small	500	Fine – even (straight grain)	Good	Non-durable	Int. joinery*
Pink to chocolate brown	Western red cedar	*Thuja plicata*	Canada USA	Small	390	Medium to coarse (straight grain)	Good	Durable	Int. and ext. joinery Cladding
White	Whitewood (European spruce)	*Picea abies*	Europe	Small	470	Fine	Good	Non-durable	Int. joinery*

Table 1.3(b) Some properties of various hardwood species (*continued on opposite page*)

Colour of heartwood	Species		Origin	(a) Moisture movement	(b) Approx. density (kg/m³)	(c) Texture (see Table 1.2)	(d) Working qualities (cutting, nailing, etc.)	(e) Durability	(f) General usage
	Common name	Latin (botanical) name							
Reddish brown	African mahogany	*Khaya* spp.	West Africa	Small	530	Medium – varies (interlocking grain)	Medium	Moderate	Int. and ext. joinery Plywood – veneer†
Reddish brown	American mahogany	*Swietenia macrophylla*	Central & S. America	Small	560	Medium	Good	Durable	Int. and ext. joinery
Light brown	Afrormosia	*Pericopsis elata*	West Africa	Small	710	Fine to medium (interlocking grain)	Medium	Very durable	Int. and ext. joinery Veneer
Light brown	Beech	*Fagus sylvatica*	Europe	Large	720	Fine (straight grain)	Good	Perishable	Int. joinery Plywood – veneer
Light brown	Elm	*Ulmus* spp.	Europe USA Japan	Medium (can vary)	580	Medium/coarse (coarse grain)	Medium	Non-durable	Int. joinery Veneer†

Table 1.3(b) Some properties of various hardwood species (*continued*)

Colour of heartwood	Species Common name	Latin (botanical) name	Origin	(a) Moisture movement	(b) Approx. density (kg/m³)	(c) Texture (see Table 1.2)	(d) Working qualities (cutting, nailing, etc.)	(e) Durability	(f) General usage
Mid to dark brown	Iroko	*Chlorophora excelsa*	East and West Africa		660	Coarse/ medium (interlocking grain)	Medium to difficult	Very durable	Int. and ext. joinery Plywood veneer
Reddish brown	Keruing	*Dipterocarpus* spp.	Malaysia (S.E. Asia)	Large	740	Medium	Difficult	Moderate	Int. and ext. joinery*
Light/ dark	Meranti Lauan	*Shorea* spp.	Malaysia Indonesia Philippines (S.E. Asia)	Small	550–710	Coarse/ medium (slight interlocking grain)	Medium	Moderate to durable	Int. and ext. joinery Plywood
Yellowish	Ramin	*Gonystylus* spp.	Indonesia (S.E. Asia)	Large	670	Medium (straight grain)	Medium	Non- durable	Int. joinery Mouldings and trim
Reddish brown	Sapele	*Entandrophragma cyndricum*	West Africa	Medium	640	Fine to medium (interlocking grain)	Medium	Moderate	Int. and ext. joinery Plywood – veneer†
Light brown	Oak	*Quercus* spp.	Europe ———— America Japan	Medium	670–790	Medium to coarse	Medium to difficult	Durable (not all species)	Int. and ext. joinery Plywood – veneer† ———— American red and Japanese unsuitable for ext. joinery†
Whitish	Sycamore	*Acer pseudoplatanus*	Europe	Medium	630	Fine (straight grain)	Medium to good	Perishable	Int. joinery Veneer
Light brown	Teak	*Tectona grandis*	Burma Thailand Indonesia	Small	660	Medium (straight grain)	Medium	Very durable	Int. and ext. joinery* Veneer†
Reddish brown	Utile	*Entandrophragma utile*	West Africa	Medium	660	Medium (interlocking grain)	Medium to good	Durable	Int. and ext. joinery Plywood – veneer

* Structural – timber suitable for situations where strength is a major factor.
† Decorative appearance – timber noted for its decorative properties.

1.2 Properties of timber

The physical properties of timber will depend on the wood species, growth characteristics, subsequent conversion, and final processing.

Table 1.3 forms the basis for selecting which of the hardwood and softwood species listed will be suited to a given situation, with regard to their

a)   moisture movement,
b)   density,
c)   texture,
d)   working qualities,
e)   durability,
f)   usage.

Moisture movement

(Column (a)) refers to the amount of movement which might affect the dimensions and shape if timber cut from that particular species is subjected to conditions liable to alter its moisture content (m.c.) after having been dried to suit its end use, i.e. in a state of equilibrium with its environment. Any increase or decrease in moisture content will result in the timber either expanding (swelling) or contracting (shrinking). Proportional amounts of change are illustrated in Fig. 1.28 of Book 1.

Density

(Column (b)) refers to the mass of wood tissue and other substances contained within a unit volume of timber. It is expressed in kilograms per cubic metre (kg/m³). It therefore follows that the m.c. of timber must affect its density – figures quoted in Table 1.3 are average for samples of 15% m.c.

Factors affecting timber density (Fig. 1.12) are

a)   the m.c. in the cell walls (bound moisture),
b)   the presence of free water in voids (free moisture),
c)   the amount of cell-wall substance in relation to air space.

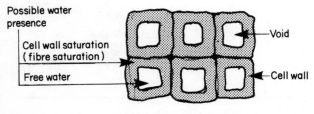

Possible water presence

Cell wall saturation (fibre saturation)

Free water

Void

Cell wall

Horizontal section through cells

Fig. 1.12   Factors affecting timber density

Because the density of solid wood tissue (1506 kg/m³) is the same for all species, the differences in density between the species listed in Table 1.3 indicate the proportion of cell tissue per unit volume.

Density is a property closely associated with the hardness of timber and its strength properties (see 'Strength' below).

Texture

(Column (c)) – as previously stated – refers to grain character and condition, which can greatly influence the use and working properties of timber.

Working qualities

(Column (d)) refers very broadly to how the timber will respond to being cut and machined, but not necessarily to how it will respond to glue. Factors which may influence these qualities include grain condition (Table 1.2); a hard, soft, or brittle nature; and the presence of any cell intrusions which could cause the wood to be of an abrasive, corrosive, or greasy nature.

Durability

(Column (e)) depends on the wood's natural resistance to fungal attack, which is influenced by moisture content and the amount of sapwood.

The standards of durability quoted refer to sample species of a 50 mm × 50 mm section of 'heartwood' left in contact with the ground for a number of years. Species quoted as 'perishable' may be regarded as having less than five years' life, whereas 'very durable' indicates more than 25 years' life.

General usage

(Column (f)) indicates the situations to which the particular species is best suited, provided the specimen is of the correct quality with regard to either strength or appearance. There are only a few species which satisfy both strength and appearance requirements, and those which do are usually very expensive – English oak being a good example.

Table 1.4 shows how cut sections of timber often reflect their end use.

Table 1.4 Relationship between timber sections and their end use

Timber	Work stage	End use
SAWN	Structural carpentry	Carcassing – skeleton framework (walls, floors, roofs)
SAWN AND PROCESSED	Carpentry – 1st fixing (work done before plastering)	Partitions, floors, decking, grounds, window boards, door casing/linings
PROCESSED BOARD AND TRIM	Carpentry – 2nd fixing (work done after plastering)	Shelving, trims (skirting boards, architrave), falsework (panelling, pipe boxes, etc.)
	Joinery (exterior)	Windows, doors, gates, etc.
PROCESSED BOARDS, SECTIONS, AND TRIMS	Joinery (interior)	Carcase units, cupboards, doors, stairs, and fitments

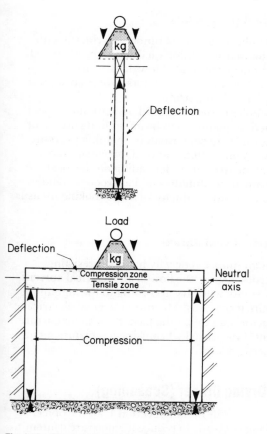

Fig. 1.13 Structural timber being subjected to a load

Other properties

Strength

Because of its high strength-to-weight ratio, timber can be suited to situations requiring either compressive or tensile strength. Figure 1.13 shows how a structural timber member can be stressed. The posts are being compressed from their ends by a beam and its load and the ground. The beam is being subjected to stresses of compression (wood tissue being pushed together) on its upper surface and tension (tissue pulled apart) on its lower surface – the neutral axis being a hypothetical dividing line where stresses are zero.

If either of the stressed zones – particularly the tension zone – is damaged or distorted either by natural defects such as knots etc. (see Book 1, Section 1.6) or by defects associated with seasoning (Section 1.4), the beam could be drastically weakened. The cutting of notches in stressed timber members must always be strictly controlled (see Fig. 7.11) – notches should not be cut into the tension zone.

Structural weakness may also be due to other factors. Examples are given in Table 1.5 portraying the following:

a) timber section in relation to load;
b) low density of timber (particularly of the same species) at an acceptable moisture content;
c) high moisture content;

d) direction of growth rings – wood cells and tissue should be positioned to provide greatest strength;

e) fast growth of softwood produces low-density wood because of the large thin-walled cells (tracheids);

f) slow growth of hardwood produces a greater number of large vascular cells (vessels) and a lesser number of strength-giving fibres.

Table 1.5 Guide to the condition of timber in relation to its general strength characteristics

Condition	Strong	Weak
a) Siting – position (stiffness)		
b) Density	High (heavy)	Low (light)
c) Moisture content	Low	High
d) Direction of growth rings	Tangential-sawn softwood	Quarter-sawn softwood
e) Fast grown	Ring-porous hardwood	Softwood
f) Slow grown	Softwood	Ring-porous hardwood

Effects of fire

Timber is, of course, combustible. The importance of this is not that timber burns but for how long it retains its structural stability while being burnt.

It would be impossible to light a camp fire with heavy logs – it is much better to start with dry twigs and then build up the size of material gradually until the fire has a good hold. Then, and only then, can the logs be added so that the fire will, provided there is enough draught and conditions are right, burn steadily for some time – probably hours without needing replenishing.

It can therefore be said that the rate at which timber burns must be related to

a) its sectional size,
b) its moisture content,
c) its density,
d) an adequate supply of air.

Provided an adequate section is used, timber will retain its strength in a fire for longer than steel or aluminium, even though they are non-combustible.

One effect flaming has on timber is to form a charcoal coating over its surface which acts as a heat insulator, thus slowing down the rate of combustion. Failure rate therefore depends on the cross-sectional area of the timber.

Flame-retardant treatments are available but, although useful for unstressed wood, their presence may have an adverse effect on the strength of timber (see Section 1.7).

Thermal properties

The cellular structure of timber provides it with good thermal-insulation qualities – species which are light in weight and of low density are particularly effective insulators, where structural strength is not important.

Examples of timber's good insulation and poor conduction qualities can be found in its use as, or in, external wall construction – helping prevent heat loss from within and cold intrusion from without, or vice versa, depending on the local environment or climate – and as wooden handles for metal (good conductor of heat) cooking utensils etc.

Electrical resistance

Although timber generally has good resisting properties against the flow of an electrical current, its resistivity will vary according to the amount of moisture it contains. Moisture is a good electrical conductor, and this is the basis on which moisture content is measured when using a moisture meter, as described later.

1.3 Drying timber (Seasoning)

As described in Book 1 (Section 1.7), the object of drying timber is to reduce its moisture content to a level acceptable for its end use (Table 1.6), thus increasing its stability, strength, durability, and

Table 1.6 Moisture contents of timber for various purposes

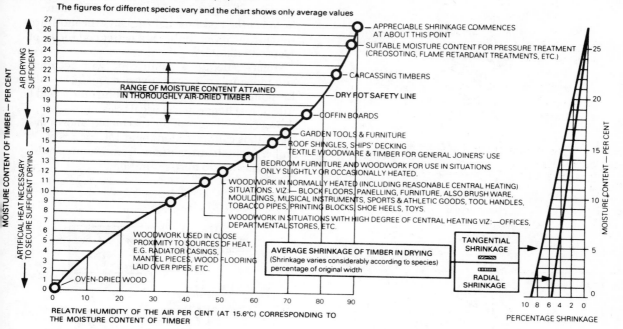

The figures for different species vary and the chart shows only average values

workability to bring them in line with those properties listed in Table 1.3.

Methods used to achieve this end are many and varied; for example

a) air drying (natural drying),
b) kiln drying (artificial drying),
c) air drying followed by kiln,
d) forced-air drying,
e) climate chambers,
f) dehumidifiers,
g) vacuum drying,

Note Methods (d) to (g) are outside the scope of this book.

Because the object of drying timber is to remove water from the cells (Fig. 1.12), moisture content is considered first.

Moisture content (m.c.)

The moisture content of timber is the measured amount of moisture within a sample of timber expressed as a percentage of the dry weight of timber within the sample. If the weight of water present exceeds that of dry timber, then moisture contents of over 100% will be obtained.

There are two methods of determining moisture content:

i) the oven-drying method,
ii) by using electrical moisture meters,

Oven-drying method (Fig. 1.14)

A small sample cut from the timber which is to be dried (see Fig. 1.23) is weighed to determine its 'initial' or 'wet' weight. It is then put into an oven with a temperature of 103°C ± 2°C until no further weight loss is recorded, its weight at this stage being known as its 'final' or 'dry' weight.

Once the 'wet' and 'dry' weights of the sample are known, its original moisture content can be determined by using the following formulae:

$$\text{moisture content (as a percentage)} =$$

$$\frac{\text{initial or wet weight} - \text{final or dry weight}}{\text{final or dry weight}} \times 100$$

or   moisture content (as a percentage) =

$$\left(\frac{\text{initial or wet weight}}{\text{final or dry weight}} - 1 \right) \times 100$$

For example, if a sample has a 'wet' weight of 25.24 g and a 'dry' weight of 19.12 g,

$$\text{moisture content} = \frac{25.24\,\text{g} - 19.12\,\text{g}}{19.12\,\text{g}} \times 100$$

$$= 32\%$$

or   moisture content $= \left(\dfrac{25.24\,\text{g}}{19.12\,\text{g}} - 1 \right) \times 100$

$$= 32\%$$

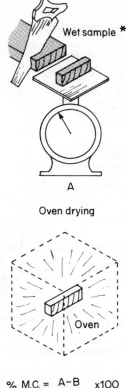

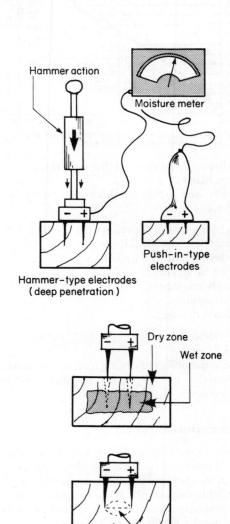

timber is converted to a moisture content which can be read on the calibrated scale of the meter – the lower the resistance, the greater the moisture content, since wet timber is a better conductor of electricity than dry.

Meters generally will only cope, with accuracy, with moisture contents between 6 and 28% – above this point (the fibre-saturation point) there will be little or no change in electrical resistance. However, the larger of the two meters shown in Fig. 1.16 is calibrated for about 150 species of timber for moisture contents from 7% (depending on the species) to fibre saturation.

$$\% \text{ M.C.} = \frac{A - B}{B} \times 100$$

Fig. 1.14 Method of determining moisture content by oven drying a timber sample

Electrical moisture meters (Fig. 1.15)

These are battery-operated instruments which usually work by relating the electrical resistance of timber to the moisture it contains.

Moisture content is measured by pushing or driving (hammer-type) two electrodes into the timber. The electrical resistance offered by the

Fig. 1.15 Battery-operated moisture meters

Fig. 1.16 Protimeter 'Mini C' (top) and 'Diagnostic Timbermaster' moisture meters in use

Moisture meters are more than just a useful aid for making spot checks – in fact in the practical sense, when used in conjunction with the timber drying procedures of air and kiln drying they can be better than the oven-drying method. Whenever a moisture meter is used it is important that:

- probes can reach the part of the timber whose moisture content is needed (depends on the sectional size of the timber and the type of instrument);

b) allowance can be made for the timber species – timber density can affect the meter's reading;

c) the temperature of the timber is known – meter readings can vary with temperature;

d) certain chemicals are not present in the timber – for example, wood preservatives or flame-retardant solutions.

Tests may be necessary when sorting large batches of timber or checking the condition of assembled or fixed carpentry and joinery, particularly if a fungal attack is in evidence or suspected – in which case a moisture meter would be invaluable.

Air drying (natural drying – see also Book 1, pages 13–14)

This method of drying is often regarded as a process which timber undergoes before being kiln-dried. This is possibly true in the main but, if space is available and time is unimportant, many species can be successfully dried by this method to between 20 to 18% moisture content when left for periods of three months to two years (depending on the species, the sectional size, and the overall drying environment).

The success of air drying will depend on the following factors:

a) weather protection,
b) site conditions,
c) stacking,
d) atmospheric conditions.

Weather protection

Except when drying certain hardwoods which can be dried as an open-piled 'boule' (the log being sawn through-and-through and then reassembled into its original form – see 'Stacking'), a roof is employed to protect the stack from direct rain or snow and extremes in temperature. Its shape is unimportant, but corrugated steel should be avoided in hot climates because of its good heat-conducting properties which would accelerate the drying process. Roof coverings containing iron are liable to rust and should not be used where species of a high tannin content (such as oak, sweet chestnut, afrormosia, Western red cedar, etc.) are being dried, otherwise iron-staining is possible where roof water has dripped on to the stack.

Shed sides may be open (see Book 1, Fig. 1.25) or slatted. Adjustable slats enable the air flow to be regulated to give greater control over the drying process. End protection can also be provided by

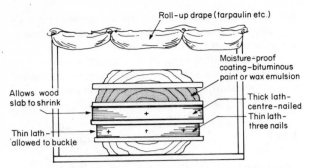

Fig. 1.17   End treatment (normal stack or boule) – helps prevent end splitting

this method – unprotected board ends are liable to split as a result of the ends drying out before the bulk of the timber, and hardwoods like oak and beech are particularly prone to this problem. Other methods used to resist this particular seasoning defect are shown in Fig. 1.17, namely:

a)   treating the end grain with a moisture-proof sealer – bituminous paint or wax emulsion, etc.;

b)   nailing laths over the end grain – thick laths should be nailed only in the middle of the board, to allow movement to take place;

c)   hanging a drape over the end of the boule or stack.

Site conditions

The whole site should be well drained, kept free from vegetation by blinding it with a covering of ash or concrete, and kept tidy. If fungal or insect attack is to be discouraged, 'short ends' and spent piling sticks should not be left lying around.

Sheds should be sited with enough room left for loading, unloading, carrying out routine checks, and other operations.

Stacking

Stacks of sawn timber can be as high as safe limits of stacking and stability will allow. Their width should not exceed 2 m, but adjacent stacks can be as close as 300 mm to each other.

Piling sticks (stickers) should never be made from hardwood, or they could leave dark marks across the boards. Their size and distance apart will vary according to board thickness, drying rate, and species. They must always be positioned vertically one above the other, otherwise boards may 'bow' as shown in Fig. 1.18(a). Stacks with boards of random length may require an extra short stick as shown in Fig. 1.18(b).

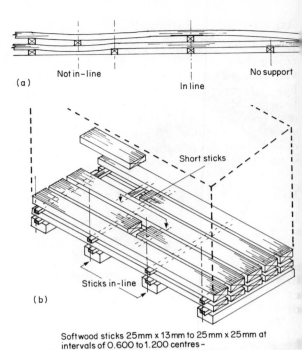

Softwood sticks 25 mm x 13 mm to 25 mm x 25 mm at intervals of 0.600 to 1.200 centres – depending on board thickness and drying rate.

Fig. 1.18   Build-up of stack

Figure 1.19 shows how boules are piled in log form. Certain hardwoods are often dried in this way, to ensure that the dried boards will match one another in colour and figure.

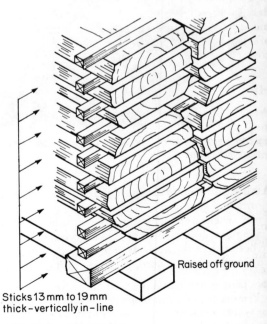

Fig. 1.19   Boules-piled in log form

Atmospheric conditions

It is impracticable to generalise on an ideal drying environment when atmospheric conditions can vary so much between seasons and countries. It is, however, important that whatever means are used to regulate the drying rate of timber should be directed at achieving uniformity throughout the whole stack – otherwise the timber could become distorted or suffer other defects due to uneven shrinkage (see 'Drying defects', Section 1.4).

Kiln drying (artificial drying)

Drying kilns are closable chambers designed to speed up the process of drying timber.

The essential features of such a kiln are

1) heat,
2) ventilation,
3) humidification,
4) air circulation.

Heat

Is often provided via steam or hot water pipes. The fuel used to fire the boiler may be of wood waste, oil, gas, or coal.

Ventilation

Is achieved by adjustable openings strategically positioned in the kiln wall or roof. Alternatively, a dehumidifier can be used to extract unwanted moisture and channel it outside the kiln in the form of water – thus conserving heat and reducing fuel costs.

Humidification

When the amount of moisture leaving the wood is insufficient to keep the humidity to the required level, jets of steam or water droplets may be introduced into the chamber.

Air circulation

Is promoted by a single large fan or a series of smaller fans, located either above or to the side of the stack, depending on the kiln type. All the above must be controlled such that the whole process can be programmed to suit the species, thickness, and condition of the wood.

Prescribed kiln schedules are available to take the wood through the various stages of moisture content (say from 'green' to 15% m.c.). These list

in descending order the appropriate kiln temperature and relative humidity needed for each stage of drying.

The temperature and the amount of water vapour in the air entering the stack are measured with a kiln hygrometer to assess the relative humidity of the air, which will determine the rate at which the wood dries. A 'wet-and-dry-bulb hygrometer' is shown in Fig. 1.20. It consists of two thermometers. The dry bulb measures air temperature. The wet bulb is attached to a fabric wick dipping into a jar of distilled water. Evaporation of water from the wick lowers the temperature of the wet bulb by an amount depending on the rate of evaporation, which in turn depends on the relative humidity of the air – the lower the humidity of the air, the greater the rate of evaporation, and hence the lower the wet-bulb temperature. The greater the difference between the wet-bulb and dry-bulb temperatures, therefore, the lower the humidity of the air.

Relative humidity at a particular temperature is expressed as a percentage (fully saturated air having a value of 100% RH). Less water vapour at the same temperature means a lower relative humidity; therefore by lowering the relative humidity, drying potential is increased. It must also be remembered that the higher the air temperature, the greater its vapour-holding capacity.

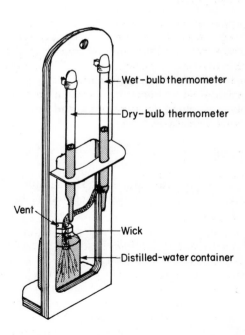

Fig. 1.20 Wet-and-dry-bulb hygrometer

Very broadly speaking, it can be said that kiln drying involves three stages:

i)   heating up the wood without it drying – low heat, high humidity;
ii)   starting and continuing drying – increased heat, less humidity;
iii) final stages of drying – high heat, slight humidity.

Kiln samples

Adjustments to obtain the next stage of drying conditions can only be made once the correct moisture content for the present stage has been reached, which is uncertain until suitable tests have been carried out. Sample testing involves the removal of several boards from different parts of the stack and weighing them – moisture content can then be determined by their loss in weight. Figure 1.21 shows how timber is piled and provision is made in a stack for the easy removal of samples. Figure 1.22 shows a sample board being removed from the stack – notice also the control panel outside the kiln with a wet-and-dry-bulb temperature recorder.

Before drying begins, small test-pieces are cut from the kiln samples (Fig. 1.23) and are immediately tested for moisture content, using the oven-drying method previously described. If the kiln sample is now weighed, to obtain its wet weight, the dry weight of the kiln sample board can be estimated:

$$\text{dry weight of kiln sample} = \frac{\text{wet weight}}{\text{m.c.}/100 + 1}$$

Once this is known, any future m.c. occurring during the drying process can be calculated by simply reweighing the kiln sample and using the following formula:

$$\text{m.c.} = \left(\frac{\text{current weight}}{\text{dry weight}} - 1 \right) \times 100\%$$

As an example, assume that a kiln sample weighs 14.5 kg. With a moisture content of 40%, its dry weight would be

$$\text{dry weight of kiln sample} = \frac{\text{wet weight}}{(\text{m.c.}/100 + 1)}$$

$$= \frac{14.5 \text{ kg}}{(40/100 + 1)}$$

$$= \frac{14.5 \text{ kg}}{1.4}$$

$$= 10.36 \text{ kg}$$

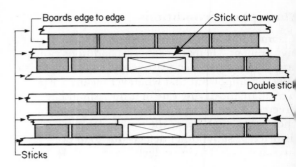

Fig. 1.21   Two examples of how provision can be made for the easy removal of kiln samples

Fig. 1.22   Inspection and removal of a kiln sample

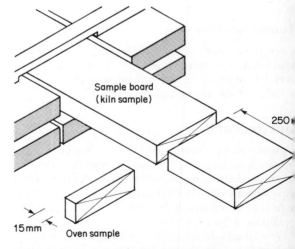

Fig. 1.23   Cutting an oven sample to determine the moisture content of the sample board

After a period of drying, it is found that the sample now weighs 12.2 kg; therefore its current m.c. would be

$$\left(\frac{\text{current weight}}{\text{dry weight}} - 1\right) \times 100\% =$$

$$\left(\frac{12.2}{10.36} - 1\right) \times 100\% = 17.8\%$$

Note   Moisture meters like the 'Diagnostic Timbermaster' (Fig. 1.16), used with a hammer probe, provide a quick and accurate means of taking moisture-content readings – as stated earlier.

Drying kilns

These can be divided into two groups: (i) compartment kilns and (ii) progressive kilns.

Compartment kilns

These are sealable drying chambers (compartments) which house batches of timber, loaded on trolleys, until their drying schedule is complete.

Figure 1.24 shows how these kilns can be arranged – separately with single, double, or triple tracks, or joined together in a row (battery).

Kilns may be sited outside, like that shown in Fig. 1.25, which has an individual steam-raising boiler (see also Book 1, Fig. 1.26), or undercover like the battery of dryers shown in Fig. 1.26 which receive heat from a central boiler plant.

Figure 1.27 shows how timber piles are stacked in a double-track kiln with a central unit containing the fan, heaters, humidifiers, ventilators, and controls. Figure 1.28 has this unit to one side to accommodate three tracks.

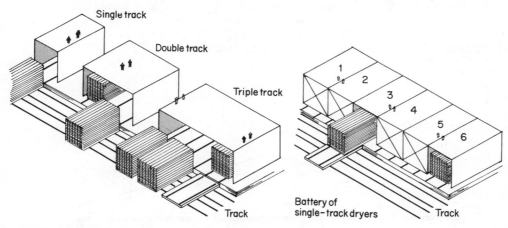

Fig. 1.24   Compartment kilns (for the sake of clarity some doors are not shown)

Fig. 1.25   Externally sited compartment dryer

Fig. 1.26   Battery of compartment dryers

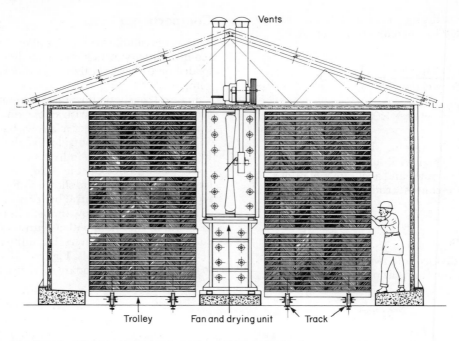

Fig. 1.27   Cross-section through a 'Wells' double-track high-stacking prefabricated timber dryer

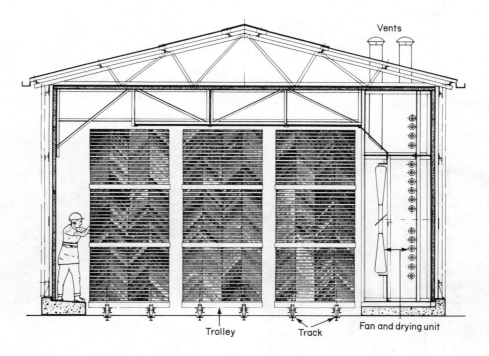

Fig. 1.28   Cross-section through a 'Wells' triple-track high-stacking prefabricated timber dryer

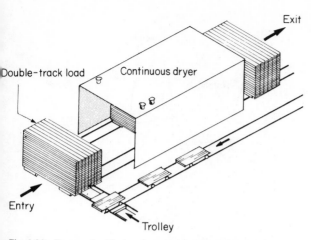

Fig. 1.29 Progressive kiln (continuous dryer) – doors not shown

Progressive kilns

In these, green timber enters the kiln at one end and – after a period of time which can be as short as three to five days, depending on the species and the cross-section – emerges from the exit at the opposite end in a much drier state. The whole process enables timber to be dried by continuous means.

Figure 1.29 shows how batches of timber are lined up on trolleys on tracks outside the tunnel, ready to follow those already inside.

On entry, each batch will go through a series of stationary drying stages, which start cool and humid but end with the last stage being warm and dry. When a batch leaves the tunnel, a new batch will enter from the other end to take its place.

1.4 Drying defects

Successful drying depends on how drying preparations are made and on how the whole operation is carried out.

Green timber is usually in a pliable state – after drying, it stiffens and sets. For example, a green twig will bend easily but, if held in that position until dry, it will set and remain partially bent. Therefore, if green timber is allowed to become distorted, either by incorrect piling or due to unbalanced shrinkage during its drying, then permanent degrading could result.

From Table 1.7, which should be used in conjunction with pages 13–14 of Book 1, it can be seen that most of these degrading defects can be attributed to both the unevenness and the speed at which moisture is removed from the wood.

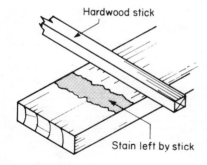

Fig. 1.30 Stick marks

Fig. 1.31 Cupping

Fig. 1.32 Diamonding

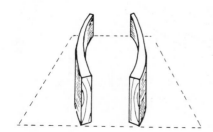

Fig. 1.33 Bow (distortion)

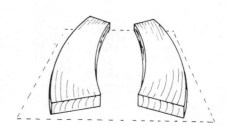

Fig. 1.34 Spring (distortion)

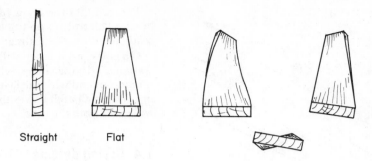

Straight   Flat

Fig. 1.35   Twist

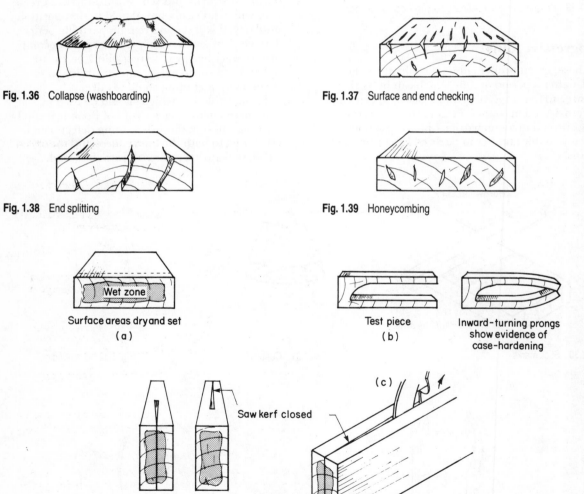

Fig. 1.36   Collapse (washboarding)

Fig. 1.37   Surface and end checking

Fig. 1.38   End splitting

Fig. 1.39   Honeycombing

Wet zone

Surface areas dry and set
(a)

Test piece
(b)

Inward-turning prongs
show evidence of
case-hardening

Saw kerf closed

(c)

Fig. 1.40   Case-hardening

Table 1.7 Analysis of drying defects

Defect term	Definition	Cause	Possible preventative measures
Stains			
Stick marks (Fig. 1.30)	Evidence of where sticks have been laid across a board	Using dirty sticks or sticks with an acid content, e.g. certain hardwoods	Use only softwood sticks – see 'Stacking'.
Sap stain (blue stain)	Bluish discolouration of the timber (fungal growth)	Close piling sapwood with over 25% m.c. – poor air circulation	Ventilate around each board
Distortion (warping)			
Cupping (Fig. 1.31)	Curvature in the cross-section	Differential shrinkage due to the position of growth rings	Use a low-temperature schedule when kiln seasoning.
Diamonding (Fig. 1.32)	Square-sectioned timber becomes diamond-shaped	Diagonally positioned growth rings induce the greatest amount of shrinkage.	As above.
Spring (Fig. 1.33)	Curvature along a board's edge	Differential shrinkage longitudinally along irregular grain	As above.
Bow (Fig. 1.34)	Curvature along a board's width	Sagging in the pile and/or differential shrinkage as above	As above.
Twist (Fig. 1.35)	Spiral deformity – propeller-shaped	Irregular grain – spiral and/or interlocking	As above.
Collapse (washboarding) (Fig. 1.36)	Buckling of the timber's surface – corrugated effect	Uneven shrinkage – drying too rapidly and collapsing the spring wood	As above.
Checking and splitting			
Checks (surface and end checks) (Fig. 1.37)	Parting of the grain, producing cracks (fissures on the surface and/or end of timber).	Surfaces drying much quicker than the core	Use high humidity in the early drying stages
Splits (end splitting) (Fig. 1.38)	As above, but cracks extend through the timber from face to face.	End grain drying quicker than the bulk of the wood	See 'Stacking', sealing end grain, etc.
Honeycombing (Fig. 1.39)	Parting of the grain internally	Shrinkage of the inner zone after outer zone has become case-hardened	Use low-temperature schedules; use high humidity early on.
Case-hardening			
Case-hardening (Fig. 1.40)	Outer zone of the timber dries and 'sets' before inner zone, setting up internal stresses between the two.	Rapid surface drying due to low humidity early on, or high temperature in the later stages of drying	High humidity early on, keeping a check on temperature throughout

Figure 1.40(a) Shows how the different zones could appear. Figure 1.40(b) shows how a test-piece can be cut to test the extent of case-hardening – releasing tension will cause it to distort as shown. The danger of cutting case-hardened material should be apparent, e.g. trapping saw blades etc. (Fig. 1.40(c)).

1.5 Wood-rotting fungi

Fungi were dealt with in general terms in Book 1, pages 16–17. Here we are mainly concerned with the two most common wood-destroying types: the 'dry-rot' fungus (*Serpula lacrymans* – formerly known as *Merulius lacrymans*) and 'cellar rot' *Coniophora puteana* – formerly *Coniophora cerebella*), which is one of the wet-rot fungi. Both dry rot and cellar rot belong to a group of rots known as 'brown rots', as wood which they attack appears darker (brown) in colour (and, on drying, becomes brittle). On the other hand, 'white rots' – which include other wet rots outside the scope of this book – lighten the colour of the wood.

Wherever the following conditions exist, fungi will inevitably become established – the type of fungus and its characteristic life style being in the main determined by the amount of dampness.

a) *Food* – in the form of cellulose from the woody tissue of sapwood and non-durable heartwood (see Section 1.2, 'Durability').

b) *Moisture* – in the first instance, wood will have had to attain a moisture content in excess of 20%.

c) *Temperature* – in general terms, fungi can survive in temperatures between blood heat of 98.6°F (37°C) and 32°F (0°C). Low temperatures may reduce their growth; high temperatures, on the other hand, will kill them.

d) *Air* – an essential requirement for the growth and respiration of fungi.

Once fungi are established, it is only a matter of time before the wood substance starts to decompose and structural breakdown occurs with the result that the wood:

a) loses its strength,

b) becomes lighter in weight,

c) changes its colour by becoming darker or lighter – depending on the type of fungus.

The following text should be read in conjunction with Table 1.8.

Dry-rot fungus (*Serpula lacrymans*)

Timber becomes liable to dry-rot attack when its moisture content exceeds 20% and it is sited in positions of poor ventilation – good ventilation being one of the prime factors for keeping timber below 20% m.c. Probably the most ideal conditions for development of dry rot are situations of high humidity with little or no ventilation and a moisture content of between 30 to 40%. Under these conditions, dry rot can spread to distant parts of the building – provided a source of moisture is available, nothing seems to stand in the way of its reaching fresh supplies of wood.

This fungus has moisture conducting strands which help it sustain growth while it travels behind plaster, or through walls in search for food.

Once dry rot becomes widely spread, its initial starting point (often its main source of moisture) becomes more difficult to find – the fungus may have travelled from room to room, from floor to floor, even into or from the roof, and have re-established itself with another moisture supply. Places which are most suspect will have evolved from bad design, bad workmanship, or lack of building maintenance. Some of these locations are shown in Fig. 1.41.

A typical outbreak of dry rot to a timber floor is shown in Figs 1.42 and 1.43.

Tell-tale signs of an outbreak could be one or more of the following features:

a) Smell – a distinct mushroom-like odour (damp and musty).

b) Distorted wood surface – warped, sunken (concave), and/or with shrinkage cracks. Tapping with a hammer produces a hollow sound, and the wood offers no resistance when pierced with a knife.

Table 1.8   General characteristics of the wood-rotting fungi: dry rot and cellar rot

	Name of fungus	
	Dry rot (*Serpula lacrymans*), Figs 1.42 to 1.44	Cellar rot (*Coniophora puteana*), Figs 1.46 and 1.47
Appearance:		
fruiting body (sporophore)	Plate or bracket, white-edged with rust-red centre (red spore dust)	Plate – not often found in buildings – olive green or brown
mycelium	Fluffy or matted, white to grey, sometimes with tinges of lilac and yellow	Rare
strands	Thick grey strands – can conduct moisture from and through masonry.	Thin brown or black strands, visible on the surface of timber and masonry
How wood is affected	Becomes dry and brittle, breaking up into large and small cubes, brown in colour.	Exposed surfaces may initially remain intact. Internal cracking, with the grain and to a lesser extent across. Dark brown in colour.
Occurrence	Within buildings, originating from damp locations	Very damp parts of a building, e.g. affected by permanent rising damp, leaking water pipes, etc.
Other remarks	Once established it can spread to other, drier, parts of the building.	Fungal attack ceases once dampness is permanently removed.

Note   Both the above fungi can render structural timbers unsafe.

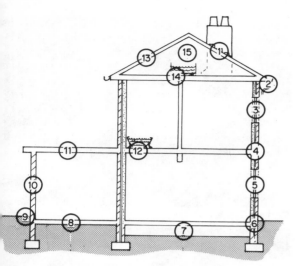

g. 1.41 Possible causes of dampness 1 – defective step
shing 2 – defective or blocked gutter/fallpipe 3 – window
ndensation – insufficient external weathering 4 – bridged cavity
all (mortar droppings etc.) 5 – defective or omitted vertical d.p.c. to
ors and/or windows 6 – defective or omitted horizontal d.p.c.;
ocked air brick/grate; ground above d.p.c. 7 – no ventilation to
der-floor space 8 – defective or omitted d.p.m. 9 – defective or
itted d.p.c; ground above d.p.c. 10 – solid wall of porous
asonry 11 – defective roof covering; unvented void (cold-deck
nstruction); defective or omitted vapour barrier 12 – defective
mbing – water spillage 13 – defective roof covering
– defective plumbing 15 – unvented roof space

Fig. 1.43 Severe attack of dry rot exposed

1.42 Surface effect of a dry-rot attack

The appearance of fruiting bodies
(sporophores) in the form of a 'plate' (skin) or
'bracket'.

Fig. 1.44 Dry-rot damage

d) The presence of fine rust-red dust, which is the
spores from a fruiting body.

Exposed unpainted timber may reveal a
covering of a soft carpet of whitish mycelium.
Established mycelium may appear as a greyish
skin with tinges of yellow and lilac. Wood already
under attack will have contracted into cuboidal
sections (Fig. 1.44).

The fruiting body of the fungus, which is capable
of producing millions of spores, appears as an
irregularly shaped rusty-red fleshy overlay with
white edges.

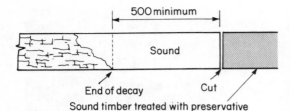

Fig. 1.45   Cutting away timber affected by rot

Eradication of dry rot

The following stages are necessary even for the smallest of outbreaks.

i)   Eliminate all obvious sources of dampness – some possible causes are shown in Fig. 1.41.

ii)   Investigate the extent of the outbreak – removing woodwork and cutting away plasterwork as necessary.

iii) Search for further causes of dampness both within and outside the areas of attack.

iv)   Remove *all* affected woodwork (timber should be cut back as shown in Fig. 1.45) and fungus from the building to where it can be safely burnt or disposed of.

v)   Sterilise and treat affected and surrounding walls, concrete floors, etc. with a suitable fungicide.

vi)   Treat with preservative all sound woodwork within and beyond the affected area.

vii) All replacement timber must be below 20% moisture content and treated with preservative (paying special attention to end grain).

Note   See also 'Preservatives' in Section 1.7.

Cellar rot *(Coniophora puteana)*

The most favourable locations for this rot are those which can permanently provide wood with a moisture content of about 40 to 50%. Examples of where suitable conditions may be found are listed below (see also Fig. 1.41):

a)   in damp cellars or other rooms below ground level;

b)   beneath leaking water pipes and radiators;

c)   behind and under sinks and baths, etc. – due to persistent overspill and splashing;

d)   areas of heavy condensation – windows, walls, roofs;

e)   areas where water can creep (by capillarity) and remain free from evaporation – behind damaged paintwork, under window sills (Fig. 1.46) and thresholds, etc.

Fig. 1.46   Typical wet-rot location

Fig. 1.47   Wet-rot damage to end grain

f)   above a defective damp-proof course (d.p.c.) or membrane (d.p.m.).

g)   in timber in permanent contact with the ground, or sited below a d.p.c. or d.p.m.

However, unlike dry rot, once the moisture source has been removed and the wood has dried out, cellar rot becomes inactive.

Figure 1.47 shows the effect of wet-rot damage on end grain.

The extent of the decay may not at first be obvious, because the outer surfaces of the timber often appear sound. It is not until these surfaces are tested with the point of a knife or bradawl that the full extent of the attack is known. The rot is, however, reasonably confined to the area of dampness.

One of the biggest dangers associated with cellar rot is that its area of confinement may eventually become taken over by dry rot, in which case a major problem may well have developed.

There are several other wet rots, all of which require reasonably high percentages of moisture. This can make formal identification difficult.

Eradication of wet rot

Unlike dry rot, the one controlling factor here is dampness. However, all wet rot should be treated in the following manner:

i) Remedy as necessary any sources or defects responsible for the dampness (Fig. 1.41).

ii) Dry out the building.

iii) Remove and safely dispose of affected timber, cutting back well into sound wood (Fig. 1.45).

iv) Treat with preservatives all remaining timber in close proximity (paying special attention to end grain).

v) *All* replacement timbers must be below 20% moisture content and treated with preservative (paying special attention to end grain).

Note Because the strands of cellar rot do not penetrate masonry, treatment is more localised.

Prevention of wood rot

Theoretically, rot should not occur in a building which has been correctly designed, built, and maintained. For prevention, all that is required is that the moisture content of wood should never be allowed to exceed 20%. The only obvious places where wood is subject to conditions likely to achieve that level are external situations – for example external joinery, e.g. doors, windows, gates, etc. – in which case, each of the items will require permanent protection against the entry of moisture, unless timber with an appropriate natural-durability rating has been used in its construction.

1.6 Attack by insects

Book 1, pages 17–18, dealt in general terms with the life cycle of woodboring beetles (insects).

This section, in conjunction with Table 1.9, is concerned with the life style of five insects found damaging wood in Britain and the termite which is a major problem in warmer climates.

Table 1.9 General characteristics of some wood-destroying insects

Name of insect	Common furniture beetle (*Anobium punctatum*), Fig. 1.48	Powderpost beetle (*Lyctus* spp.), Fig. 1.49	Death-watch beetle (*Xestobium rufovillosum*), Fig. 1.50	House longhorn beetle (*Hylotrupes bajulus*), Fig. 1.51	Wood-boring weevils (*Pentathrum* or *Euophryum* spp.), Fig. 1.52
Adult characteristics:					
size	3 to 5 mm	5 mm	6 to 8 mm	25 mm	3 to 5 mm
colour	Reddish to blackish brown	Reddish brown to black	Chocolate brown	Grey/black/brown	Reddish brown to black
flight	May to August	May to September	March to June	July to September	Any time
Where eggs laid on or in wood	Crevices, cracks, flight holes, etc.	Vessels of large-pored hardwoods	Fissures* in decayed wood, flight holes	Fissures* in softwood, sapwood	Usually on decayed wood
Size of larvae	Up to 6 mm	Up to 6 mm	Up to 8 mm	Up to 30 mm	3 to 5 mm
Bore dust (frass)	Slightly gritty ellipsoidal pellets	Very fine powder, soft and silky	Bun-shaped pellets	Cylindrical pellets	Very small pellets
Dia. of flight (exit) hole	2 mm	0.75 to 1.5 mm	3 mm	6 to 9 mm Ellipsoidal (oval)	Up to 1.5 mm
Life cycle	2 years or more	10 months to 2 years	3 to 15 years	3 to 11 years	7 to 9 months
Wood attacked:					
type	SW and HW, plywood with natural adhesive	HW, e.g. oak, elm, obeche	HW, especially oak	SW	SW and HW, plywood with natural adhesive
condition	Mainly sapwood	Sapwood	Previously decayed by fungus	Sapwood first, heartwood later	Usually damp or decayed
location	Furniture, structural timbers etc.	Timber and plywood while drying or in storage	Roofs of old buildings, e.g. churches	Roofs and structural timbers	Area affected by fungus (usually wet rot) – cellars etc.
Other remarks	Accounts for about 75% of woodworm damage in UK. Resin-bonded plywood and fibreboard immune.	Softwood and heartwood immune	Attacks started in HW can spread to SW	Restricted to areas of Surrey and Berkshire in UK.	Attack continues after wood has dried out.

* Fissures – cracks, narrow openings, and crevices

Actual size approx. 3 to 5 mm long

Flight holes

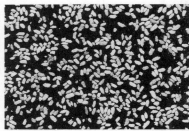

Bore dust

Fig. 1.48   Common furniture beetle (*Anobium punctatum*)

Actual size approx. 5 to 6 mm long

Fig. 1.49   Powderpost beetle (*Lyctus* spp.)

Severe internal damage and apparently superficial external damage

Bore dust

Actual size approx. 6 mm long

Fig. 1.50   Death-watch beetle (*Xestobium rufovillosum*)

Typical damage

Bore dust

Actual size approx. 25 mm long

Fig. 1.51   House longhorn beetle (*Hylotrupes bajulus*)

Typical damage to rafters

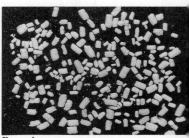

Bore dust

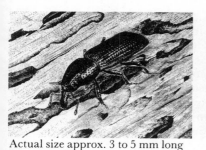

Actual size approx. 3 to 5 mm long

Typical internal damage

Bore dust

Fig. 1.52 Wood-boring weevils (*Pentathrum huttoni* and *Euophryum confine*)

Common furniture beetle *(Anobium punctatum)*, Fig. 1.48

In Britain, the larvae (woodworm) of this beetle are responsible for about 75% of the damage caused by insects to property and contents – attacking both softwood and temperate hardwood, but with a preference for sapwood.

The adults range from reddish to blackish brown in colour, 3 to 5 mm in length, and live for about 30 days. Eggs, about 80 in number, are laid by the female in small fissures or old flight holes.

The eggs hatch in about four weeks. Larvae, up to 6 mm long, may then burrow through the wood for over two years, leaving in their trail a bore dust (frass) of minute ellipsoidal pellets resembling fine sand. The larvae finally come to rest in small chambers just below the surface of the timber, where they pupate (change into chrysalides).

Once the transformation to beetle form is complete – between May and August – they bite their way out, leaving holes about 2 mm in diameter known as 'flight' or 'exit' holes. After mating, the females will lay their eggs to complete their life cycle.

Because of these beetles' ability to fly, very few timbers are exempt from attack. Some West African hardwoods do, however, seem to be either immune or very resistant to attack – they are:

abura African walnut afrormoisia
idigbo iroko sapele

Evidence of an attack is provided by the flight holes and, in many cases, the bore dust ejected from them. Old and neglected parts of property, including outbuildings, provide an ideal habitat for these insects – attics and cellars are of particular interest to them, as they may remain undisturbed there for many years.

Powderpost beetle *(Lyctus spp.)*, Fig. 1.49

There are many species of 'Lyctus' – as denoted by 'spp.' – the most common of which is *Lyctus brunneus*, to which this text applies. 'Powder-post' refers to the very fine powdered bore dust (frass) left by the larva of the Lyctus beetle.

This beetle attacks the sapwood of certain hardwoods such as ash, elm, oak, walnut, obeche, and ramin, doing so when they are in a partially dried state and at their most vulnerable, for example stacked in a timber yard. Infestation can be passed into the home with newly acquired hardwood furniture or recently fixed panelling or block floors etc.

The life cycle of this insect can be very short, which means that infestation can spread rapidly.

Wood species with small pores (which restrict egg laying) or insufficient starch (essential for larval growth) are safe from attack. Softwoods, some hardwoods, and heartwood are therefore immune.

Death-watch beetle *(Xestobium rufovillosum)*, Fig. 1.50

This is aptly named because of the ticking or tapping noise it makes during its mating season, from March to June, and its often eerie presence in the rafters of ancient buildings and churches.

Its appearance is similar to that of the common furniture beetle, but similarities end there. Not only is it twice as long, it generally only eats hardwoods which have previously been attacked by fungi – and it seems to have a preference for oak. The duration of the attack may be as short as one year or as long as ten or more – much depends on the condition of the wood, the amount of fungal decay, and environmental conditions.

Because the life cycle can be lengthy, infestation over wide areas is limited.

House longhorn beetle *(Hylotrupes bajulus)*, Fig. 1.51

This large beetle takes its name from its long feelers. It is a very serious pest on the mainland of Europe and in parts of Britain – mainly in north Surrey and certain areas of adjoining counties – where it attacks the sapwood of seasoned softwood.

Because of the size of its larvae and of the subsequent bore holes left by extensive tunnelling over a number of years, the extent of the damage caused to structural timbers has led to much concern in Britain – so much so that it is mandatory under the Building Regulations to treat with a wood preservative all softwood used for the purpose of constructing a roof or ceiling within those geographical areas stated in the regulations (mainly Surrey and Berkshire).

Wood-boring weevils *(Pentarthrum huttoni* and *Euophryum confine)*, Fig. 1.52

Beetles in this group have distinct protruding snouts from which their feelers project. The two species which concern us here are *Pentarthrum huttoni* and *Euophryum confine* (known also as the New Zealand weevil). Their attacks are usually confined to the damp and/or decayed (usually wet rot) sapwood and occasionally heartwood of both softwood and hardwood.

Unlike the other wood boring beetles mentioned, both the adult and larvae of these insects carry out 'boring' activities. Tunnels tend to follow the pattern of the wood's grain.

General eradication of insects

Both obvious and suspected areas of activity should be fully investigated to determine

a) the extent and nature of the attack,
b) the size and shape of flight holes,
c) the amount and nature of bore dust (frass),
d) the moisture content of the wood,
e) if fungal attack is in evidence,
f) the species and nature of the wood under attack.

It should then be possible to establish which beetle or beetles are responsible for the attack.

False alarms can occur if flight holes are the only symptom, as an attack may have taken place before use, when the timber was 'green' (not dried) – in which case, the culprits would have been killed during timber drying. For example, the 'ambrosia beetle' takes its nourishment from a fungus which forms in the bore holes, and this fungus cannot survive in wood with a moisture content lower than 35% – the beetle and its activity are therefore eliminated once the timber has been dried, but evidence of its attack remains in the shape of very small pin-holes, usually with a dark stain around their perimeter (covering tunnel walls).

However, assuming the outbreak being investigated is active, the following measures should be taken:

i)   Where possible, open up affected areas – cut away wood which is badly attacked and carefully remove it from the site and dispose of it in a safe manner.

ii) If load-bearing timbers are affected, seek professional advice about their structural stability. Repair or replacement of such items usually involves propping and shoring, to safeguard against structural movement or collapse.

iii) Thoroughly clear the whole area of debris. Dust removal may involve the use of an industrial vacuum cleaner.

iv) Treat remaining and replacement timbers with an approved preservative (insecticide). Application should be carried out in accordance with the maker's directions – preservative is usually applied with a brush, and particular attention should be given to end grain, fissures, and joints.

Operatives must wear an approved face-mask, for protection from inhaling vapours given off, and protective clothing, goggles, and gloves must be used, thus protecting eyes and skin from coming into contact with the materials. Precautions must also be taken against the possibility of a fire, as, at the time of application, organic-solvent types of preservative give off a vapour which is a fire risk.

Termites

Termites are a major problem mainly in the tropics and subtropical regions of the world – they do not appear in Britain. These insects are capable of totally destroying timber at an alarming rate.

The life-style of termites is somewhat similar to that of ants; for instance, they live in colonies containing workers who carry out the destruction and soldiers who guard and protect the colony from intruders, while the kings and queens are responsible for populating the colony. Similarities, however, end there.

There are two groups of termites:

i) those which dwell entirely inside the timber they are destroying;
ii) those which dwell in the ground, in earth cavities, or under mounds of earth, away from their food supply. Access to this food supply is gained by the termites either tunnelling under or building a tunnel over any obstacle they may meet, in order to avoid the light and detection.

Prevention

Some timbers are resistant to termite attack – for example, afrormosia, iroko, and teak (there are others) – but most timbers can be given good protection by impregnating them with preservative.

Where ground-dwelling termites are a problem, buildings can be designed with physical barriers against access, such as metal or concrete caps between timber and ground, and protruding perimeter sills (shields) of metal or concrete. All building perimeters must be kept clear of shrubbery etc., which might act as a bridge the termites could use to bypass the perimeter shield.

Eradication

Methods of eradication will depend on the termite group, i.e. wood- or earth-dwelling.

With wood-dwelling types, treatment will be similar to that for other wood-boring insects, but with earth-dwelling types the problem is finding the point of entry into the building. One solution is to trench all round the building and treat the ground with chemicals – those termites then in the building will be cut off from their nest.

1.7 Special treatment of timber

Except where timber is used in its natural state for practical or economical reasons, it is generally treated with either paint, water-repellent stains, preservatives, or special solutions to either reduce or retard the effects of

a) weathering,
b) moisture movement,
c) fire;

or to eliminate the risk of

i) fungal attack,
ii) insect attack.

Paints

Exterior-quality paints and varnishes protect timber from

a) the entry of water,
b) abrasive particles,
c) solar radiation (not varnishes).

The weathering effect of wind, rain, and sunlight will gradually degrade the surface of unpainted timber by breaking down its surface. Water penetration of non-durable wood is the greatest problem – not only will it result in variable dimensional change (expansion on wetting, contraction on drying), it also increases the risk of fungal attack. Sunlight causes the wood to change colour and contributes to its degradation.

Correctly painted surfaces should give relatively good protection for up to five years, provided that

i) the timber is throughly dry (i.e. has the appropriate moisture content) before paints are applied,
ii) the surfaces have been prepared correctly,
iii) the paint manufacturer's recommended number of coats are given,
iv) the end grain has received particular attention with regard to coverage,
v) inspections are carried out regularly.

If a painted surface becomes damaged, moisture may enter the wood and become trapped behind the remaining film of paint, and further paint failure and wood decay could quickly follow.

Water-repellent exterior stains

Exterior stains provide a clear or coloured, matt or semi-gloss, water-repellent surface which – unlike paint – allows the wood grain to show through. They give protection against weathering and fungal staining (because of their wood-preservative content) for up to four years or more, depending on the stain quality.

One of the main advantages over paint (with the exception of micropore paints, see Book 1, page 16) is that they allow the wood to breathe – thereby allowing trapped moisture to escape, and avoiding the problems previously stated.

Preservatives

Unless treated with a preservative, sapwood and non-durable heartwood will be liable to attack by insects and, if the moisture content is above 20%, by fungi.

Wood preservatives are solutions which are either applied or introduced into the wood to make it toxic (poisonous) to insects and fungi.

There are three preservative groups:

i)   organic-solvent types,
ii) water-borne types,
iii) tar-oil types.

Organic-solvent preservatives

These use a medium of organic solvents to transmit the toxic chemicals into the wood. After application the solvents evaporate, leaving the wood toxic to fungi and insects.

Methods of application include brushing, spraying, immersion, and certain pressure processes. The solvents used are generally volatile and flammable, and extreme care must therefore be taken at the time of application and in storing containers.

These preservatives do not affect the dimensions of timber (cause swelling) or have a corrosive effect on metals. The ability to glue or paint timber is unaltered after treatment once the preservative has dried.

Water-borne preservatives

These use water to convey the toxic chemicals. There are two types: one for treating unseasoned (green) timber and the other for seasoned timber. They do, however, have to be applied by special means to ensure thorough penetration – they are not suitable for brush application.

These preservatives are non-flammable and may be painted over when dry. Drying is always necessary after treatment.

Tar-oil preservatives

These are derived from coal tar and are ideal for preserving exterior work which is not to be painted. They do not usually have any corrosive effect on metals, but they will stain most porous materials they contact.

The most common form of tar-oil preservative is 'creosote', which is light to dark brown in colour and can be applied by various processes including brushing and spraying. It has a strong odour for some time after its application.

Methods of application of preservatives

Preservatives are applied by one of two general methods: pressure or non-pressure. Figure 1.54 gives some indication as to how the method used affects the depth of penetration.

Pressure methods

Timber is put into a sealable chamber into which preservative is introduced under pressure as described below:

High Pressure Methods:
i)   Full-cell process
ii) Empty-cell process

Low Pressure Method:
i)   Double-vacuum process

Full-cell process   After sealing the chamber, the air is removed by using a vacuum pump. With the chamber still under a vacuum and after a prescribed period, preservative is introduced, filling the chamber. The chamber will remain filled until the timber has absorbed sufficient preservative, which will vary according to the permeability of the wood.

On completion, surplus preservative is pumped out of the chamber back into its storage tank. Finally, a further vacuum is set up in the chamber, but only of sufficient strength to remove any surface preservative – cell cavities will remain full.

This process normally uses creosote and water-borne preservatives.

Empty-cell process   In this case, timber in the chamber is subjected to air pressure, either by introducing compressed air before the preservative (in this case creosote) or as a result of the preservative being introduced unter pressure.

After the wood has absorbed sufficient preservative, the pressure is released and surplus preservative is driven out of the wood by the expanding air in the wood cells. A vacuum is then used to draw off any residue. Although the wood cells are emptied, their walls remain fully treated.

Double-vacuum process   (A typical treatment plant is shown in Fig. 1.53(a) and (b)). The chamber is sealed and a partial vacuum is created. The chamber is then filled with preservative and pressurised to atmospheric pressure, or above, depending on the process. After a prescribed period, it is then drained and a final vacuum is created to remove excess preservative from the timber.

This process is mainly used to apply organic-solvent preservatives into timber for exterior joinery – required depths of penetration are therefore less than for the full-cell and empty-cell processes but better than for immersion. It does not cause timber to swell or distort, and all machining should be done before treatment.

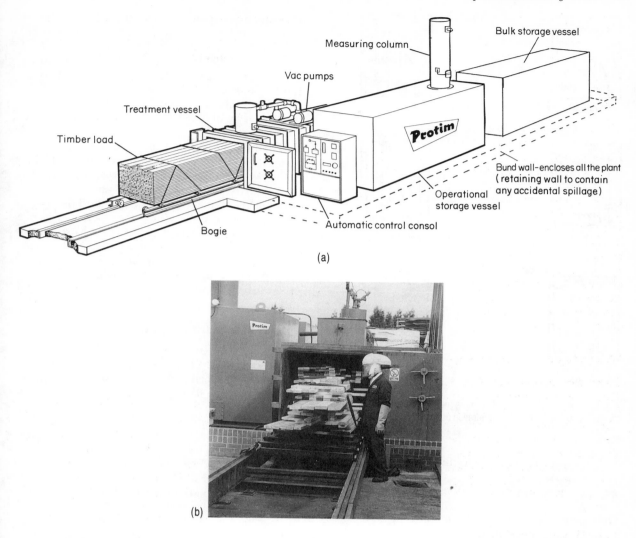

Fig. 1.53 'Protim' prevac double vacuum treatment plant

Non-pressure methods

With the exception of the 'hot and cold open-tank treatment' and 'diffusion' methods described below, depth of preservative penetration with non-pressure methods is often limited to just below the surface of the wood – as shown in Fig. 1.54.

Brushing Can be used for applying creosote and organic types of preservative but, because of low penetration, it is not a suitable method for timbers which come into contact with the ground. As a rule, retreatment is advisable every three to four years.

Spraying Similar penetration and conditions apply as for brushing.

Because of the health risk associated with applying preservatives, precautions should always be taken – particularly when spraying – to ensure that

a)   only coarse sprays are used, to avoid atomisation;
b)   work areas are well ventilated;
c)   operatives are suitably clothed;
d)   hands are protected by gloves;
e)   mouth and nose are protected by an approved face-mask;

Sample

Hot-and-cold treatment

Brushing and spraying

Pressure

Steeping

Diffusion

Fig. 1.54   A guide to preservative impregnation in relation to different methods of application (note: penetration depth exaggerated)

f)   eyes are protected with snug-fitting goggles – not glasses;

g)   manufacturer's instructions are followed.

Deluging   In deluging, the timber is passed through a tunnel of jets which spray it with preservative. In the main, organic-solvent preservatives are used in this process.

Immersion (dipping)   The timber is submerged in a tank of preservative (coal-tar oils or organic-solvent types) for a short period, then allowed to drain.

Steeping   The timber is submerged for periods ranging from a few hours to weeks, depending on the wood species, the sectional size of the timber, and its end use. Steeping is a suitable method for preserving fence posts etc.

Hot and cold open-tank treatment   The timber is submerged in a tank of preservative (coal-tar oil) which is heated. It is then allowed to cool in the tank or is transferred to a tank of cold preservative. This treatment is suitable only for permeable timbers and sapwood.

Coal-tar oils are flammable, therefore extra care is necessary with regard to the heat source.

Diffusion   This method of treatment is only associated with freshly felled green timber which, at the saw mills, is immersed in a water-borne preservative (usually boron salts) and then close-piled and placed under cover until the preservative has diffused into the wood.

The type of water-borne preservative used in this process is liable to leach out from the wood, which makes the timber unsuitable in wet locations unless an impervious surface treatment is given, i.e. paint or varnish.

Flame-retardant treatments

By impregnating the timber with a solution of various salts or other chemicals, or coating its surface with special paints, it is possible to reduce the rate at which flame would normally spread over its surface.

Timber impregnated with flame-retarding salts is not normally suited to exterior use, because the salts are liable to leach out. Strength properties of timber will be reduced as a result of this treatment.

By using a special process, leaching can be avoided and strength properties of the timber remain unaltered.

Paints and varnishes

Some paints and varnishes are 'intumescent' – they swell when subjected to heat and protect the wood by forming an insulating layer over the surface of the timber. Others give off a gas which protects against flaming.

Other treatments

Other treatments applied to timber include mould oils and release agents – see Section 10.1 on formwork.

2

Portable powered hand tools

This chapter deals with portable powered hand tools associated with the following forms of motive power:

a) mains electricity and electricity derived from portable generating sets (electric powered hand tools),

b) batteries (cordless tools),

c) compressed air (pneumatic tools),

d) explosive powder (cartridge-operated fixing tools).

2.1 Electrically powered hand tools

Book 1, Chapter 4, dealt with the electric drill, screwdriver, belt sander, and orbital sander, together with the basic principles of supplying electricity to the tool.

There is, however, one other method of providing a supply of electricity – i.e. a petrol- or diesel-driven portable generator – which could increase the versatility of these tools even further by not having to rely on a 'mains' supply.

Figure 2.1 shows a typical portable petrol-driven generator capable of producing 2.4 kW, with dual outlets for either 120 or 240 volts supply.

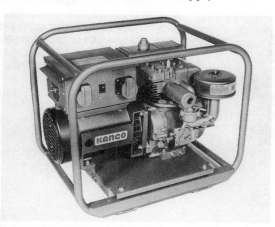

Fig. 2.1   Petrol-driven portable generating set

Circular saws

Figure 2.2 shows a typical portable circular saw. Its main components have been identified by name and/or function. Saws like this are capable of cutting a variety of materials, such as

a) timber:
    i) softwood,
    ii) hardwood;

b) manufactured boards:
    i) plywood,
    ii) particle board (chipboard),
    iii) blockboard and laminboard,
    iv) fibre board – hardboard and insulation board;

c) laminates;

d) synthetics;

Provided the correct blade is used, a very satisfactory cut can be made. Table 2.1 can be used as a general guide to blade suitability.

The efficiency of the blade, and to some extent the safety of the cutting operation, will depend on the sharpness of the blade. Blades must therefore be kept sharp. Blade replacement must be carried out according to the tool maker's instruction, as this process may differ slightly between makes. Blades cut on the upward stroke and rotate in the direction of the arrow – usually shown on the top guard.

Portable circular saws are capable of carrying out the following operations:

i) *Ripping*   Figure 2.3 shows how the same ripping guide is used and how, by using a temporary fence, wide boards can be sawn.

ii) *Rebating*   Two cuts are required, as shown in Fig. 2.4.

iii) *Cross-cutting*   Accurate cuts can be made by using a jig like the one shown in Fig. 2.5(a).

iv) *Cutting bevels* – **(a)**   with the grain (Fig. 2.6) any angle between 90° and 45°;

Table 2.1 General guide to circular-saw blades

Blade type	Suitable for cutting	Operation	Remarks
Combination blade	Softwood and hardwood	Ripping and cross-cutting	General-purpose
Cross-cut blade	Softwood and hardwood	Cross-cutting	Fine finish
Composition and wallboard blade	Composition boards and aluminium sheet	—	Fine-toothed blade
Flooring blade	Reclaimed timber etc.	Capable of cutting the occasional nail	Specially tempered blade
Planer and mitre blade	Softwood and hardwood	Ripping and cross-cutting	Produces a very smooth finish
Tungsten-carbide-tipped blades	Softwood and hardwood, manufactured boards, laminates and synthetics, etc.	Ripping and cross-cutting	Sawn finish depends on teeth formation and pitch*

*See Section 3.2, Figs 3.6 and 3.7.

(b) across the grain (Fig. 2.7) any angle between 90° and 45° or by modifying the jig shown in Fig. 2.5.

v) *Cutting compound bevels*   Use a temporary fence, or a jig like that shown in Fig. 2.5 and angle adjustment. The cut resembles face and edge cuts to a jack rafter (see Fig. 9.5).

vi) *Cutting mitres*   Use a temporary fence or a jig and angle adjustment.

vii) *Cutting plough grooves*   Make a series of cuts with the grain, using the aids mentioned.

viii) *Trenching (housing)*   As in (vii), but across the grain.

With operations (i) to (vi) the depth of cut should be such that the gullets (see Figs 3.6 and 3.7) of the teeth protrude just below the under surface of the material being cut, as shown in Fig. 2.8. In this way the maximum number of teeth are in contact with the material at the same time, which also reduces the angle of cut and the tendency for the blade to break away the material as it cuts through the upper surface – the material is therefore better positioned face down.

Fig. 2.2 'Kango' portable circular saw
A – cable and cable sleeve   B – back handle and trigger switch   C – specification plate   D – motor casing   E – front handle   F – graduated quadrant (angle-cutting adjustment)   G – sole-plate   H – cut-line guide   I – adjustable graduated ripping guide (fence)   J – retractable (telescopic) saw-blade guard   K – saw blade   L – saw-blade rotation indicator   M – top guard   N – quadrant (depth-of-cut adjustment)

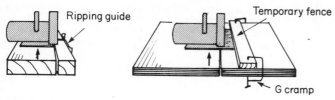

Fig. 2.3 Ripping operations

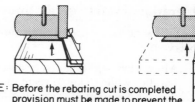

NOTE: Before the rebating cut is completed provision must be made to prevent the waste portion being shot forward due to the forward motion of the saw blade

Fig. 2.4 Rebating

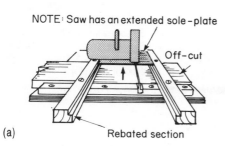

(a)

Fig. 2.5 (a) Cross-cutting jig; (b) Cross-cutting

(b)

Fig. 2.6 Bevel cutting with the grain

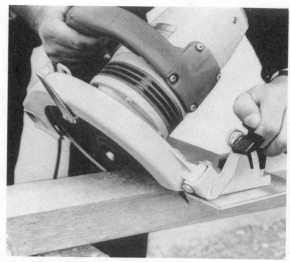

Fig. 2.7 Bevel cross-cutting

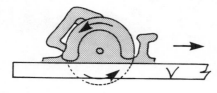

Fig. 2.8 Sawing action

Operational safety guide-lines (circular saw)

The following safety precautions should be observed:

a) Always keep both hands on the machine handles while the blade is in motion.
b) Always ensure that the spring-loaded telescopic guard is operational at all times.
c) Always remove the plug from the electricity supply before changing a blade or making any adjustments, and when the saw is not in use.
d) Ensure that the blade is fitted correctly to cut from bottom to top.
e) Never use a blunt blade.
f) Never pull the machine back towards the body after a cut – always follow through.

g) Always ensure that the workpiece is supported and held firmly.

See also the general safety rules in Section 2.5.

Jigsaws (Fig. 2.9)

Saws which fall into this category cut by the reciprocating (backward and forward or upward and downward movement) action of their blade.

Although capable of cutting a straight line – usually with the aid of either a parallel guide attachment or a temporary fence (Fig. 2.3) – the main function of jigsaws is cutting slots, curves, and circles, as shown in Fig. 2.10. Circles can be cut by using an attachment like the one shown in

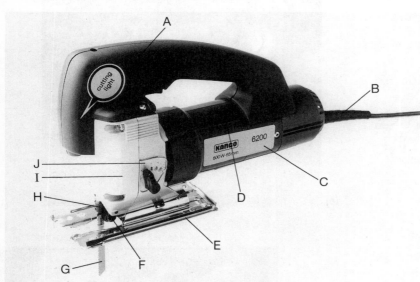

Fig. 2.9 'Kango' jigsaw A – handle and trigger switch B – cable and cable sleeve C – specification plates D – motor frame E – sole-plate F – thrust roller G – blade H – blade holder I – gearbox cover J – lever for varying degree of pendulum action (i.e. 0–111)

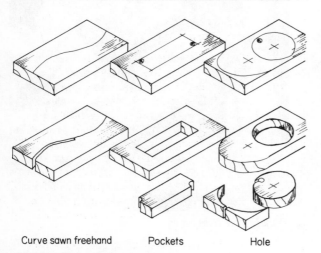

Curve sawn freehand   Pockets   Hole

Fig. 2.10 Cutting processes

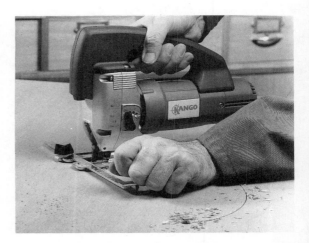

Fig. 2.11 'Kango' jigsaw with circle-cutting guide

Fig. 2.11. A rip fence is used for making parallel cuts.

Some models incorporate in their design a mechanism which directs the blade into an orbital path (pendulum action). This additional facility can be used to improve the cutting action of the blade when sawing wood, by allowing the blade to clear sawdust from the kerf on its non-cutting stroke, thereby reducing unnecessary friction.

Figure 2.12 gives an impression of how the orbital action operates. The degree of orbiting will depend on the material being cut – as indicated, straight reciprocating action is used for metals, and the amount of orbiting is increased to suit the degree of softness of the material being cut – the softer the wood, the greater the orbit.

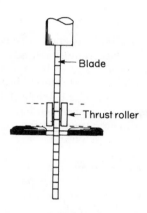

Front elevation

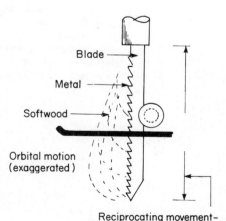

Reciprocating movement–
blade cuts on the upward stroke

Side elevation

Fig. 2.12 Orbiting (pendulum) action of the saw blade

Provided the correct blade is used, a wide range of materials can be cut. Only those blades made by the manufacturer of the tool should be used, and a specification for blade selection is issued with each tool. Table 2.2 gives some idea of the kind of performance to be expected from a blade, but it must be stressed that this is only a guide – specifications may vary between makes.

Table 2.2 General guide to jigsaw blade suitability

Material to be cut	Max. thickness of cut*	Teeth per 25 mm	Tooth pitch mm
Softwood and hardwood	30 mm	10	2.5 mm
Softwood and hardwood	60 mm	6	4 mm
†Mild steel	3 to 6 mm	8	3 mm
PVC and acrylics	13 mm	8	3 mm
Aluminium	16 mm	8	3 mm
†Mild steel and aluminium,	1.5 to 4 mm	20	1.2 mm
Plywood and hardwood	16 mm	20	1.2 mm
†Mild steel	3 to 6 mm	12	2 mm
Aluminium	16 mm	12	2 mm
Plywood and hardwood	32 mm	12	2 mm
†Stainless steel	1mm (18 s.w.g.)	32	0.7 mm

*Details may vary between makes of tool.
† Special blades.

Cutting operation

After ensuring that the workpiece is fully supported and held firm, select the correct blade, speed (high speeds for wood, low for metals), and motion (straight or orbiting (pendulum) action). Depending on the type of work being carried out, cutting can start from one edge or after inserting the blade through a pre-bored hole, as shown in Figs 2.10 and 2.11. It is not advisable to make a 'plunge cut' (Fig. 2.13, as is sometimes practised in soft materials, because of the possibility of machine 'kickback' as the blade first makes contact with the surface. The method may also result in the blade being damaged or broken.

Apart from the general safety precautions which must be followed at all times, operatives, must be

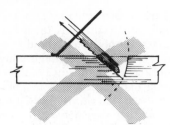

Plunge cut – <u>not</u> recommended

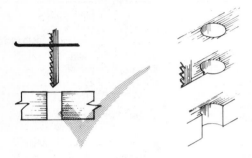

Drop into pre – bored hole

Fig. 2.13   Starting a cut midway

constantly aware of the unguarded blade –
particularly that portion which protrudes below
the saw cut.

Planers (Fig. 2.14)

Planers use a rotary cutter block, similar to a large
planing machine, thereby producing chippings as
opposed to shavings. This makes them especially
suitable for removing waste wood quickly – cuts of
up to 3 mm deep with one pass can be achieved
with the larger models. If a fine finish is required,
simply decrease the depth of cut by turning the
front hand preselect dial and move the planer over
the surface at a rate which is sufficient to prevent a
build-up of chippings at the chipping outlet.

   Apart from surfacing and rebating both
softwood and hardwood, some models are capable
of safely cutting bevels and chamfers. Most tackle
awkward and end grain with ease.

Operating

The machine must always be held with both hands
when the cutters are in motion, and it should be
applied to the firmly held and supported workpiece
only when it is running at full revolutions. Once in
contact, the 'toe-to-heel' principle should be
applied along its run. The front handle should be
held lightly, allowing the machine to be pushed by
a firmly gripped switch handle. Figure 2.15 shows
a planer in use.

   After switching the machine off, always allow the
cutters to stop revolving before putting the
machine down in the rest position. Protect the
cutters at all times by making sure they are kept

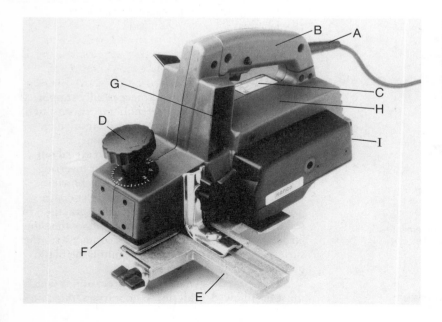

Fig. 2.14   'Kango' portable planer   A – cable
and cable sleeve   B – back handle and
trigger switch   C – specification plate
D – front handle – depth of cut adjustment
E – adjustable fence   F – sole (toe)
G – chip chute (exhaust)   H – motor housing
I – sole (heel)

Fig. 2.15 Planer in use

clear of any surface or obstacle. It is very important that the blades are kept sharp, otherwise the motor will suffer from overloading and the finish obtained will be poor.

Cutters and moving parts must never be touched for any reason whatsoever until the machine has been unplugged and completely disconnected from its power supply.

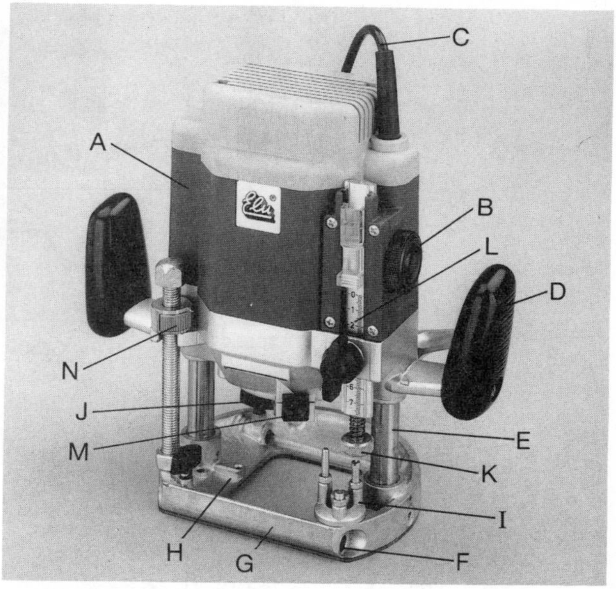

Fig. 2.16 'Elu' plunge router   A – motor housing   B – depth stop
C – cable   D – side handle   E – plunge guide bars   F – hole for
fence guide   G – baseplate   H – template guide housing   I – swivel
stop block (3 depth settings)   J – collect chuck (bit holder)   K – stop
pole and fixing knob   L – scale   M – spindle lock   N – adjustable
return stop

Routers

A router consists of a cutter rotating at between 18 000 and 24 000 rev/min, being driven by a vertically mounted motor set in a flat-bottomed framework.

Figure 2.16 shows an Elu 'plunge' router which, unlike fixed-frame types, allows the cutter to be plunged vertically into the workpiece and retracted on completion of its work. This means that the cutter need never be unduly exposed before or after the operation, thereby reducing the risk of damaging the cutter and of accidents.

Provided a suitable cutter is used, and in some cases special attachments, the following operations should be possible:

a)   cutting grooves,
b)   cutting rebates,
c)   cutting slots,
d)   cutting beads,
e)   recessing,
f)   moulding,
g)   dovetailing,
h)   trimming.

Router bits and cutters

For general wood-cutting these are made of high-speed steel, but a longer cutting life will be obtained by using bits and cutters which are carbide-tipped. Tungsten carbide (one of the hardest materials made by man) is bonded to an alloy-steel body to form a cutting edge capable of handling abrasive materials like plywood (hard glue lines), particle boards, and plastics. A few bits in the small range are available in solid carbide.

Table 2.3 in conjunction with Figs 2.17(a) to (n) will give you some idea of the kind of work which these bits and cutters are capable of.

Edge-forming bits usually include in their design a 'pilot' which allows the cutter to follow a predetermined path – either straight or curved. Pilots can be in the form of either a 'guide pin' or a 'guide roller'. With the guide-pin type of bit, the cut is restrained by the pin rubbing against the side of the wood, but movement is much easier with a roller, because of its ball bearing and lower friction.

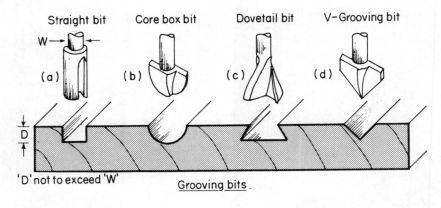

Straight bit   Core box bit   Dovetail bit   V–Grooving bit

(a)   (b)   (c)   (d)

'D' not to exceed 'W'

Grooving bits

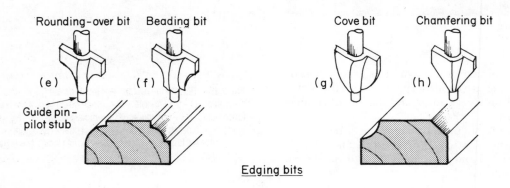

Rounding–over bit   Beading bit   Cove bit   Chamfering bit

(e)   (f)   (g)   (h)

Guide pin – pilot stub

Edging bits

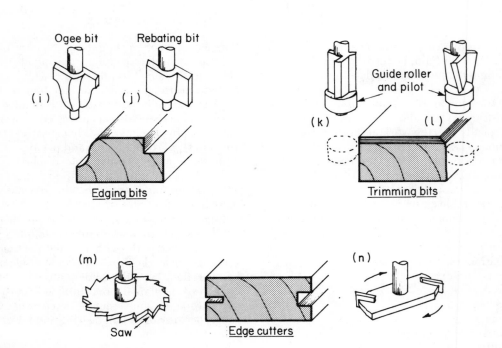

Ogee bit   Rebating bit

(i)   (j)

Edging bits

Guide roller and pilot

(k)   (l)

Trimming bits

(m)   Saw   Edge cutters   (n)

Fig. 2.17 Router bits and cutters

Table 2.3 Guide to router bits and cutters

Bit/cutter type	Operation/cutting	Mould/work profile	Figure 2.17
Grooving bits	Ploughing (housing/trenching)		(a)
	Round core-box-bit groove		(b)
	Dovetail groove		(c)
	'V' groove		(d)
Edge-forming bits	Rounding over		(e)
	Beading (ovalo)		(f)
	Coving		(g)
	Chamfering		(h)
	Ogee		(i)
	Rebating		(j)
Trimming	Square edge (flush)		(k)
	Chamfer (bevel)		(l)
Edge cutter	Sawing		(m)
	Slotting		(n)

Standard accessories

May usually include the following:

a) a guide holder with
   i) a straight fence for making parallel cuts (Fig. 2.18) and
   ii) a roller or similar attachments for curved work (Fig. 2.18).
b) a template guide for reproducing identical cuts or shapes – examples are shown in Fig. 2.19;
c) router bits (Fig. 2.17);
d) tools – spanners etc.

Operating the router

Having ensured that the workpiece is held securely, the router can now be picked up (holding it firmly with both hands) and positioned on or against the workpiece – depending on the type of machine (plunge or fixed-frame) and the work being done. Make sure that the bit/cutter is free to rotate, then start the motor. Allow it to reach maximum revolutions before making the first trial cut (usually in the waste wood or on a separate test-piece).

Deep cuts should be made by a series of shallow cuts – never cut deeper than the bit is wide (Fig. 2.17(a)), otherwise too much strain will be put on both the bit and the motor.

The cut is made from '*left*' to '*right*', to allow the bit/cutter to cut the material against its turning circle (see Fig. 2.20). Move the router just quickly enough to make a continuous cut – never overload the motor (listen to its drone) nor put unnecessary strain on the bit by pushing too hard (fast). Moving the router along too gently (slowly) will allow the cutting edge and the wood to generate heat (by friction) which could result in the bit/cutter 'blueing' (overheating) and wood being burnt. A dull (blunt) bit/cutter will have the same effect.

At the end of each operation, switch off the motor. The bit/cutter should be free of the work and allowed to stop revolving before the machine is left.

Note Throughout all routing operations there is always a risk of flying particles entering the eyes. For this reason, goggles or safety glasses must be worn by the operator.

Extra accessories

The following accessories are particularly useful for small firms, where specialised machines would be either too expensive or impracticable.

a) *Trammel point and arm* – used for cutting circles.
b) *Dovetail kit* – for producing dovetail joints.
c) *Bench stand* – converts a router into a small yet practical spindle moulder, leaving the hands free to feed the work over the table. A push stick (Fig. 3.10) and cutter guards will be required. Figure 2.21 shows an Elu table router stand in use with its fence and top and side shaw guards (which hold work against the fence and table) in position forming a tunnel guard.
d) *Trimming attachment* – used to cut veneered edging flush and/or bevelled.

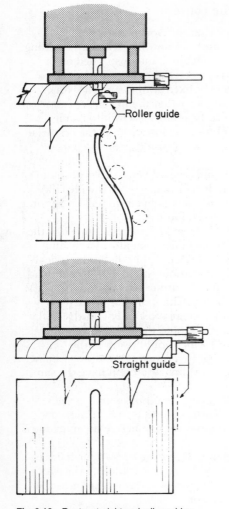

Fig. 2.18 Router straight and roller guides

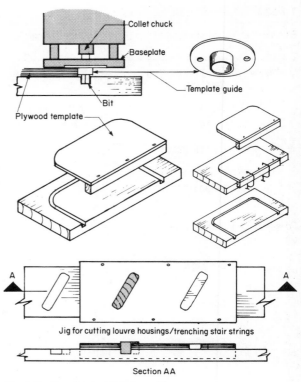

Fig. 2.19 Template and template guide

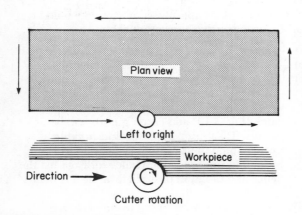

Fig. 2.20 Direction of edge cut

Fig. 2.21 'Elu' router table in use

Edge trimmers

Machines specially designed to trim over-hanging edges of veneer. Figure 2.22 shows an Elu edge trimmer being used to trim the edge of a laminate veneer – notice how the large base plate provides good surface control.

Staple guns

Hand and hammer staplers are well known for the quick effective way they can tack thin fabric and plastics to timber, or wood based boards. Probably less known are the heavy duty electronic types of staple gun – like the one shown in Fig. 2.23, which is capable of driving narrow crown staples of up to 22 mm in length. There must be many instances where such a machine can reduce labour intensive operations, such as nailing thin panel material to timber sub-frames etc.

2.2 Cordless tools (battery-operated)

These tools are powered by a rechargeable battery pack, which is located in the handle. When the power is exhausted, the pack is removed to be recharged and a freshly charged pack can replace it. The length of time that a charge will last depends on the amount of use.

When using the correct fast charger, such as the one produced by Makita (or other large power tool manufacturers), the recharging time is generally only one hour.

Makita have three separate battery systems, 7.2V, 9.6V and 10.8V which are supplied with the charger when buying the tool.

Figure 2.24 shows a power pack (battery cartridge) and charger. The charger will require a supply of electricity.

Note It is important that the maker's charging procedures – issued with each tool – are followed precisely as instructed.

Figure 2.25 shows a cordless screwdriver/drill – a typical range of specifications would be:

Chuck capacity	: 10 mm
Speed	: Single speed 600 rev/min
	: 2-Speed 250 rev/min and 600 rev/min
	: 2-Speed variable 0–250 rev/min and 0–1400 rev/min
Weight (inc battery)	: 1.1 kg to 1.7 kg
Batteries	: 7.2V, 9.6V, 10.8V
Double insulated, and reversible gearing	

Figure 2.26 shows a small cordless circular saw. Larger sizes are available with a maximum depth of cut of up to 55 mm at 90 Degrees and 36 mm at 45 Degrees.

Fig. 2.22 'Elu' trimmer

Fig. 2.23 'Rapid 137' heavy duty electronic staple gun

Fig. 2.24 'Makita' charger and power pack

Fig. 2.25 'Makita' screwdriver/drill

2.3 Pneumatic tools

Pneumatic tools derive their energy from compressed air. The compressor which produces the air may be in the form of a permanent fixture sited in the works or a mobile unit – either can vary in size depending on the number of power-line outlets and/or its mobility.

A pneumatic system will require a number of components, namely:

a) a compressor – driven by electricity or diesel;
b) a cooler (the process of compression generates heat);
c) an air receiver;
d) a water separator (when the air is cooled, water condenses out);
e) branch pipes and air-line valves;
f) a filter (moisture trap);
g) a pressure regulator; ⎫ air preparation
h) an oil lubricator; ⎭
i) air hose and couplings;
j) the tool.

A typical compressed-air installation is shown in Fig. 2.27. The air-preparation section consists of

i) a *filter* which removes remaining particles of moisture (condensation) from the air line;
ii) a *pressure regulator* – the air pressure can be pre-set and almost constantly maintained at that pressure;
iii) a *lubricator*, which allows a registered amount of lubrication to reach the tool.

Figure 2.28 shows an air-preparation unit housed in a carrying case, together with a drill and a screwdriver.

Fig. 2.26 'Makita' circular saw

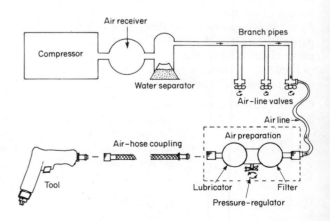

Fig. 2.27 Compressed-air installation

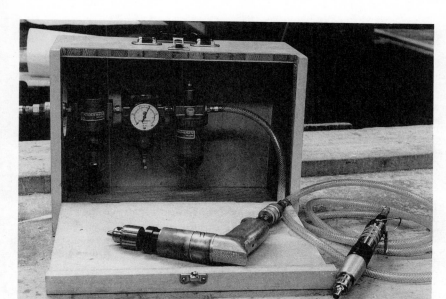

Fig. 2.28 Air-preparation unit

Tools

Pneumatic tools are available to carry out the following functions:

a) drilling,
b) screwdriving,
c) sanding,
d) circular sawing, } rotary tools

Fig. 2.29 A selection of pneumatic tools

e) tacking, stapling, and nailing.

A selection of these tools is shown in Fig. 2.29.

Rotary tools

(Types (a) to (d) above) are powered by an air motor which drives a spindle either directly or indirectly via a series of gears. A typical motor assembly for an air drill, including the 'vane' mechanism, is shown in Fig. 2.30.

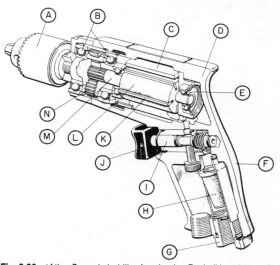

Fig. 2.30 'Atlas Copco' air drill A – chuck B – ball bearings C – cylinder D – casing E – ball bearing F – handle G – adaptor (air intake) H – filter I – valve J – trigger K – vane L – rotor M – shaft N – gear wheel

Fig. 2.31 Pneumatic fastener (nailer) in use

a)   Fooling around with air jets or apparatus must never be allowed.
b)   Never allow compressed air to be used to blow dust or grit etc. from machines – flying particles are potentially dangerous.
c)   Ensure that the air supply is at the correct pressure for the tool.
d)   Always use the correct hose size, and avoid kinks, abrasive surfaces, or corrosive material which might weaken or damage the hose.
e)   Make sure the tool is fully operational before and after use.
f)   Ensure that all operatives have read and understood the maker's instructions for both the supply and the tool.
g)   All operatives should have received adequate training in safe use of the air supply, tools, and ancillary equipment.

The Power Fastening Association publishes a code of practice for safe use of portable air-actuated fastener-driving tools.

Air tacker/stapler

These tools are capable of systematically driving a range of different fasteners with effortless accuracy; for example,

a)   small and large staples, up to 65 mm long;
b)   tacks;
c)   corrugated fasteners;
d)   round-headed nails up to 130 mm long;
e)   improved nails (twisted and annular shank);
f)   plastics sockets.

Figure 2.31 shows one of these tools in operation. The operating air pressure will depend on the size and length of the fastener and the hardness of the base material it is to enter.

Tools must be disconnected from their air line and the magazine be emptied after their work is complete and before any adjustments are made to the tool.

Safety

The absence of electricity at the tool end of the system might make one think – falsely – that dangers from the power supply have been eliminated in these tools. Unfortunately, a jet of compressed air can be very dangerous and is powerful enough to inflict serious injury to the body – if it enters the bloodstream, the result could be fatal.

Strict discipline is necessary to ensure a safe working environment; for example:

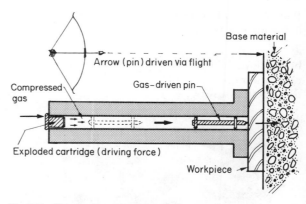

Fig. 2.32 Direct-acting tools (high-velocity types)

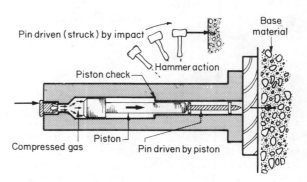

Fig. 2.33 Indirect-acting tools

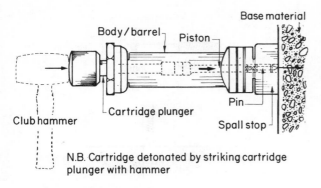

Fig. 2.34 Hammer-actuated cartridge fixing tools

Fig. 2.35 'Hilti' DX 450 cartridge tool with power regulator

2.4 Cartridge-operated fixing tools

These tools use compressed gas from an exploded charge, which had been contained in a cartridge, to propel and drive a fixing device into a base material – possibly via a workpiece. Basically there are two types: (i) direct-acting tools and (ii) indirect-acting tools.

Direct-acting tools (Fig. 2.32)

These use the contained expanding gas to drive a fixing device out of the open end of the barrel at a high velocity. They can therefore be classed as high-velocity tools.

Indirect-acting tools (Fig. 2.33)

These have a piston which intervenes between the explosive charge (cartridge) and the fixing device. There are three methods of operation:

  i) the piston is forced by the expanding gas down the barrel to strike the fixing device;
  ii) the piston and fixing device are both forced down the barrel together;
  iii) the piston and fixing device are driven together while in contact with the work face (Fig. 2.34).

The style of the tool generally resembles a pistol, as shown in Fig. 2.35 with a firing mechanism which consists of a spring-loaded firing pin operated by a trigger. Alternatively, Fig. 2.34 shows a cartridge fixing tool which is simple both to operate and to maintain but is actuated by striking a plunger with a club hammer (supplied by the manufacturer).

  The pistol type is associated with both direct and indirect types of tool, whereas the hammer-operated tool is of the indirect type and is classed as a 'low-power' tool.

  Cartridge-operated fixing tools are divided into two groups with regard to their power:

  i) low power,
  ii) high power.

The velocity of the projectile from a low-power tool must not exceed 98.5 m/s, otherwise the tool would be classed as high-power.

Fixing devices

Fixings are expected to penetrate hard materials such as mild steel and concrete, and their design and physical properties have to be matched to a specific make or model of tool. Therefore only those fixings made by the manufacturer of the tool should be used – they are not interchangeable, unless specified by the manufacturer. Failure to observe this rule could have devastating results – for example, the fixing device could shatter on impact with a hard surface.

  Figure 2.36 shows typical drive pins and how such pins are used to fix timber to concrete and mild steel. Figure 2.37 shows how 'threaded studs' with the aid of a nut and washer can provide either a permanent or a detachable fixing. Figure 2.38 shows an eyelet nail – a useful medium for hanging or suspending fitments etc.

  Various shank lengths are available to suit both the base material and the accompanying fixture – i.e. the thickness of the timber or the fixing plate etc. Shank washers or caps (depending on the make) are used to hold the fixing device in the tool and act as a guide to ensure alignment when being driven.

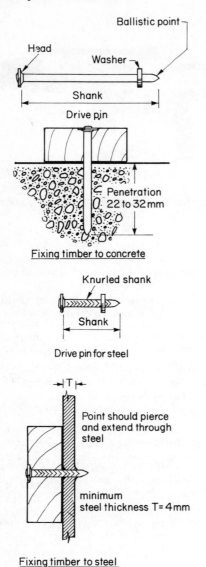

Fig. 2.36   Drive pins. Note: Penetration depths quoted are only a general guide.

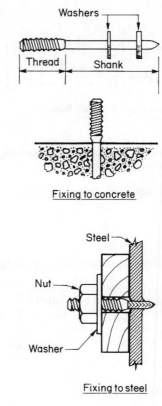

Fig. 2.37   Threaded studs

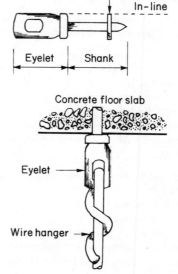

Fig. 2.38   Eyelet nails

Cartridges

Cartridges contain an explosive charge of a strength to suit different fixing situations. The actual strength of the charge is denoted by a colour and number code (Fig. 2.39). Table 2.4 lists the codes as set down in BS 4078 : Part 2 : 1989. It must, however, be noted that American and European colour codes can differ; therefore it is vitally important that the coding used must relate to the make and/or model of tool being used. Also, although cartridges may appear similar, as shown in Fig. 2.39, they are made to fit and suit a specific make or model of tool and *they are not interchangeable.*

Table 2.4 Colour-coding of cartridges to BS 4078 : Part 2 : 1989

Cartridge strength	Code numbers	Identification colour
Weakest	1	Grey
	2	Brown
	3	Green
	4	Yellow
	5	Blue
	6	Red
Strongest	7	Black

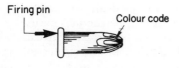

Single rim fire cartridge

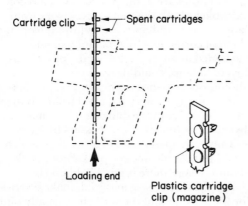

Fig. 2.39 Cartridges

Some tools have an integral (built-in) power-control system, enabling a high-strength charge to be reduced to a low-power output simply by rotating a knurled wheel (Figs 2.35 and 2.48).

Cartridges are supplied as single shots or, as shown in Fig. 2.39, assembled into a plastics cartridge clip (magazine). Cartridges should never be carried loose – only in the maker's containers (package).

Containers should indicate their contents by

a) colour-coding the label – to show the relevant cartridge strength,
b) stating the manufacturer's name,
c) giving the cartridge size,
d) stating the tool the cartridges can be used in,
e) stating the quantity of cartridges,
f) giving the batch identification number, and
g) displaying the words 'safety cartridge'.

Base materials

Suitable base materials, into which the fixing device is shot, are those which allow the pin of the device to penetrate them yet, once the pin is embedded to the correct depth, withdrawal is very difficult. Generally, good concrete, and mild steel offer very good holding properties.

Unsuitable base materials are those which are

a) *too hard* (Figs 2.40(a) and (b)) – for example, hardened brickwork rock, marble, and hardened or cast steel – any of which could cause the fixing to deflect (ricochet) or shatter;
b) *too brittle* (Fig. 2.40(c)) – for example, glass blocks, slate, and glazed and hardened tiles could shatter or break;
c) *too soft* (Fig. 2.40(d)) – for example, lightweight building blocks, some brickwork, plaster, or wood-based products could permit through penetration.

Base materials can be tested for suitability by hammering (with an engineer's hammer) a drive pin into the material as shown in Fig. 2.41. If it is found that the point of the pin becomes blunt, then the base material is too hard. If the base material starts to crack, it is too brittle; if it allows the pin to sink easily below the surface it must be regarded as too soft.

If the pin fails to penetrate to a depth of 1.6 mm or any of the above conditions are in evidence, attempts should not be made to make a fixing with a cartridge-operated tool. In doubtful situations and with brickwork, always consult the manufacturers for their advice – they will be found to be most helpful.

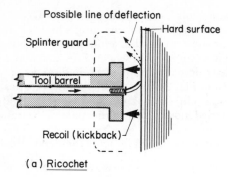

(a) Ricochet

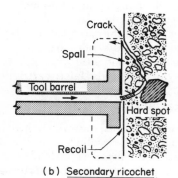

(b) Secondary ricochet

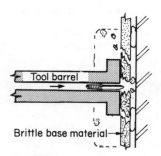

(c) Crack and/or shatter

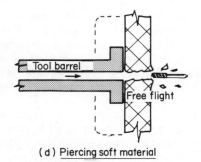

(d) Piercing soft material

Fig. 2.40   Unsuitable base materials (very dangerous practice)

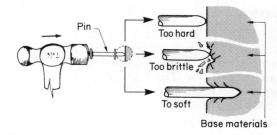

Fig. 2.41   Testing for a suitable base material

The operative

Before a cartridge tool can be used, the operative

a)   must be over the age of 18 years.

b)   must not be colour-blind – so as to avoid using the wrong strength of charge (cartridge).

c)   should have been trained in its use by the manufacturer's training instructor and have obtained a certificate of proof.
   Note   Because there are differences between makes and models of tool, each often requires a separate certificate of safe use.

d)   must be competent in its use and aware of any potential hazards which might arise, and be capable of acting accordingly – for example, a 'misfire' must be dealt with as stated in the manufacturer's instructions for use.

e)   should be suitably dressed, wearing a safety helmet, goggles, and ear protection when working in confined spaces.

f)   should be aware of the dangers of working near flammable vapour, or in an explosive atmosphere.

g)   should be aware of the possibility of a recoil (kickback) from the tool – never work from ladders, as both hands are required to operate the tool correctly. Good balance should be sustained at all times during the operations – loss of balance could prove fatal.

h)   should never load the tool away from where it is to be fired, or carry it loaded from job to job.

i)   must be aware of the dangers which may befall bystanders, onlookers, or passers by – for example, the possibility of a ricochet, which could put people in close proximity at risk.

j)   should, where there is any likelihood of a fixing piercing a base material, make provision for the area behind the wall etc. (usually blind to the operator) to be screened off so as to totally eliminate any risk of injury from either a fixing or a projected particle of the base material.

Note The manufacturer's instructions must be observed at all times. It is important, therefore, that they are always available to the operative (kept with every tool) and that they are understood – ambiguities should be clarified by either the manufacturer or an appointed agent.

Use of cartridge-operated fixing tools

Choosing the correct cartridge strength

Provided the cartridge is matched to the make and model of tool, the first trial fixing should be made using the weakest strength of cartridge, after which the strength can be increased until a suitable fixing is achieved, as shown in Fig. 2.42.

Fixing to concrete

Nails and studs driven into concrete generate a very high pressure temperature up to 900°C around the point. As a result the concrete and steel sinter (unite) firmly, producing holding powers of up to 15 kN. However, for this reaction to work

satisfactorily, the following points must be observed:

i) Establish the suitability of the base material, as previously discussed.
ii) Ensure a suitable depth of penetration (Fig. 2.36).
iii) Use the correct cartridge strength (Fig. 2.42).
iv) Always leave a minimum gap of 80 mm between fixings (Fig. 2.43).
v) Ensure that the base material is thick enough to give a satisfactory penetration – as shown in Figure 2.44.
vi) Never fix nearer than 80 mm to the edge of a crater caused by surface lifting ('spall' – Fig. 2.45).
vii) Never fix closer than 80 mm from a free edge or joint (Fig. 2.46) – Fig. 2.47 shows minimum requirements.
viii) Never fix into mortar joints.

Note Always use a splinter (safety) guard.

Never attempt fixing into pre-stressed concrete.

Figure 2.48 shows a low-velocity indirect-acting tool in use; notice that the operative is wearing safety goggles and ear protection.

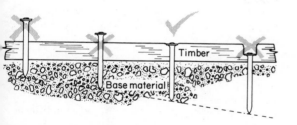

Fig. 2.42 Examples of how cartridge strength may affect penetration of the base material

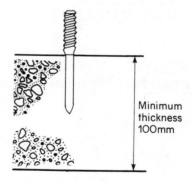

Fig. 2.44 Minimum thickness of concrete

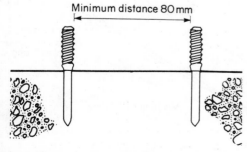

Fig. 2.43 Minimum distance between fixings into concrete

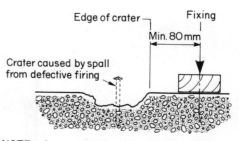

NOTE: – Spall-reduction adaptors are available

Fig. 2.45 Proximity of fixing to crater edge

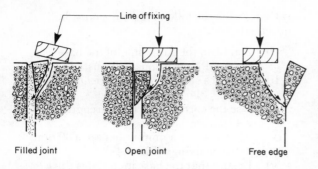

Fig. 2.46   Possible line of deflection to a joint or free edge of concrete

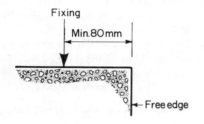

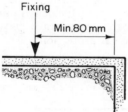

Fig. 2.47   Proximity of fixing to concrete joint or free edge

Fig. 2.48   'Hilti' low-velocity indirect-acting tool in use

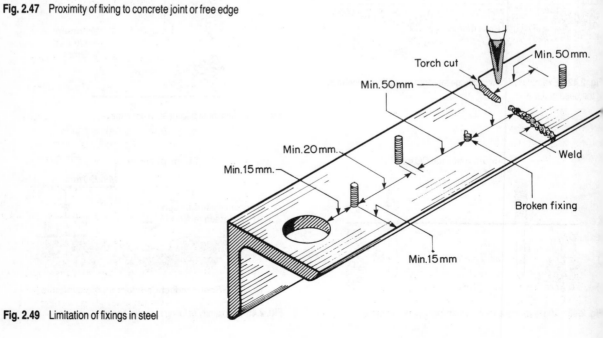

Fig. 2.49   Limitation of fixings in steel

Fixing into steel (usually mild steel and structural steel sections)

Fixings driven into steel should, where possible, pierce the steel as shown in Fig. 2.36 for maximum holding power. Fixing situations without through penetration will require special attention, as the reactive compressive forces around the pin may tend to push the fixing back out of the steel.

When fixing into steel, the following points must be observed:

i) The tool manufacturer should be consulted if there is any doubt as to the suitability of the steel as a base material.
ii) Never drive a fixing closer than 15 mm to a free edge or hole (Fig. 2.49).
iii) The distance between fixings should be not less than 20 mm (Fig. 2.49).
iv) Never drive a fixing closer than 50 mm to either a broken-off pin, a weld, or where the steel has been cut by a torch (Fig. 2.49).

Servicing

To keep the tool in good working condition, it will require regular cleaning and servicing in accordance with the manufacturer's instructions.

Any tool found to be defective must be removed from service and returned to the person responsible for its safe keeping and maintenance, clearly stating the nature of the fault, so that it can be clearly labelled accordingly before being returned to its maker or an appointed agent for repair.

Security

Cartridge-operated tools and their accessories should always be kept with the maker's case, and under the control of authorised operatives. After use, they should be kept unloaded in a locked container.

Checks on case contents should be made on collection from the store and on their return, by the storekeeper.

Correctly packaged cartridges should be kept in a compartment in the tool case – never loose.

References and recommended further reading

a) British Standard BS 4078:1966/1987, 'Cartridge-operated fixing tools'.
b) Health and Safety Executive guidance note PM 14, 'Safety in the use of cartridge-operated fixing tools'.

c) Powder Actuated Systems Association, *Guide to basic training*.

2.5 General operational safety rules for all portable powered hand tools

These rules are intended to supplement those already stated for each machine.

a) Always request *permission* to use the tool.
b) Read the maker's *operating manual* before use, and establish a clear understanding of its content. If in doubt, ask!
c) Make sure that the *tool's voltage* matches the power supply – check the specification plate (Book 1, page 93).
d) Always check the tool, its attachments, blade or cutter, cable, plug, socket, and extension lead (if used) for any *visible defect* at the start of each work period. Defective equipment must be taken out of service.
e) Make sure that the *workpiece* is securely held before starting work.
f) Maintain a good *balanced footing* throughout the whole work process.
g) Always use correctly adjusted *safety guards* where applicable.
h) Keep *work areas* clear – operative and tool must never be impeded (obstructed).
i) Never wear items of *clothing and jewellery* which could become caught in moving parts.
j) *Hair* should be short or tied back, to avoid obscuring vision and being caught in moving parts.
k) Always allow the tool to reach *maximum revolutions* before making a cut.
l) Tools must not be *laid down* until moving parts have stopped.
m) Never use tools in an *explosive atmosphere* or near *flammable substances*.
n) Keep *cables* and *hoses* clear of cutting edges and abrasive or corrosive materials.
o) Always keep blades and cutters *sharp*.
p) Ensure that blades, cutters, and bits are *correctly mounted* and *securely fixed*.
q) Always remove *adjusting keys*, spanners, etc. from the tool and place them into their holders before charging or connecting the tool to its power supply.
r) Never force a tool beyond the rate at which it was designed to work.
s) Never use electric tools in *wet or damp conditions*.
t) *Protect your eyes* – always use goggles or safety glasses (depending on the job situation – see Book 1, page 96).

u) *Face/dust-masks* should be worn where there is a dust hazard.

v) *Hands* must never be in the cutting area.

w) Always *disconnect the tool* from its power supply before changing or adjusting blades, cutters, bits, etc. or making any adjustment to guards or attachments – also whenever the tool is not in use.

x) Before using, never handle a charged or plugged-in tool with a *finger on the trigger* (tool switch).

y) Be sure the tool is *switched 'off'* before connecting it to a power supply.

Tools must be regularly maintained by the person appointed to do so by the employer.

Repairs and specialised maintenance recommended by the maker must only be undertaken by the maker or his appointed agent.

3

Woodworking machines

Book 1 (Chapter 5) dealt with the recognition and basic function of the machines which the carpenter and joiner is likely to use. It should therefore be used as a reference in conjunction with this chapter, which in the main is concerned with the application of those machines in the work situation and the relevance of the Woodworking Machines Regulations 1974 – printed in full at the end of this book.

Also included in this chapter is a basic introduction to the tenoning machine and the vertical-spindle moulding machine, as without these a machine shop would have very limited capabilities.

Machine coverage therefore includes

) cross-cutting and dimension-sawing machines,
) circular-sawing machines,
) narrow band-sawing machines,
) planing machines,
) mortising machines,
) tenoning machines,
) vertical-spindle moulding machines.

1 Cross-cutting and dimension sawing machines

(See Book 1, Section 5.1.)

This category includes both those machines which use a travelling (rolling) table to carry mber towards the saw blade and those which perate by pulling the saw blade and motor wards the timber while it remains stationary on e table.

ross-cutting and trenching machines

gure 3.1 shows a square cross cut being made th a radial-arm machine. Figure 3.2 shows just a w of the many different processes that can be rried out on this type of machine when the blade

is either accompanied or replaced by different machining heads.

Positioning timber to be cut

The full length of the timber must always be fully supported at both ends, to avoid tipping once the cut has been made. Sawn ends must never be allowed to interfere with the saw blade – Fig. 3.3 shows how this can be avoided.

'Bowed' boards (Fig. 3.3(a)) should be placed round side down, with their crown in contact with the table over the saw-cut line, and packing should be used to prevent the raised portion creating a see-saw effect with the possibility of it rocking on to the blade. If the board had been placed round side uppermost, cut ends would have been likely to drop and trap the saw blade – a very dangerous condition. Similarly when dealing with 'sprung' boards (Fig. 3.3(b)), contact with the fence at the saw-cut line is very important, otherwise the direction of saw-blade rotation could drive the cut ends of the board back towards the fence, trapping the saw blade.

Note Because the operative may use one hand to hold timber against the fence and the other to pull the saw, he must be constantly aware of the danger of a hand (particularly a thumb) coming in line with the saw cut.

Cross-cutting on a dimension sawbench

Figure 3.4 shows a dimension saw with a travelling (rolling) table being used in conjunction with a cross-cutting fence to make square (Fig. 3.4(a)), angular (Fig. 3.4(b)), and (by tilting the saw blade) compound angle (Fig. 3.4(c)) cuts.

3.2 Circular-sawing machines

(See Book 1, Section 5.2, and the Woodworking Machines Regulations Part III.)

Although 'circular saws' refers in general terms

to machines which 'divide wood with a circular blade', for the purpose of this chapter and the Woodworking Machines Regulations (Reg. 2(2)), cross-cutting saws which operate by moving the saw blade towards the material being cut are excluded.

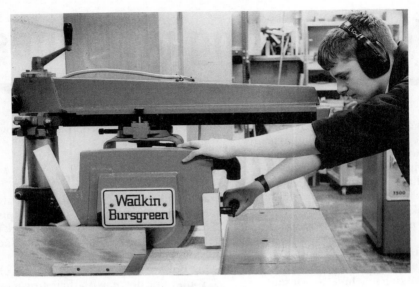

Fig. 3.1   Trainee under instruction in the use of the radial arm cross-cutting saw

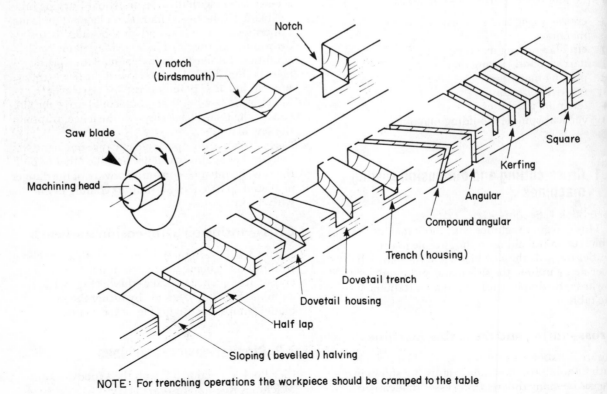

NOTE: For trenching operations the workpiece should be cramped to the table

Fig. 3.2   The versatility of the cross-cutting and trenching machine

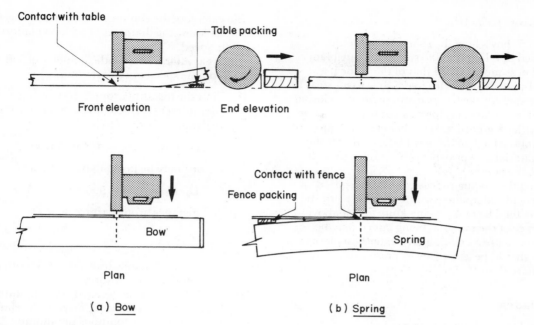

Fig. 3.3 Cross-cutting distorted timber

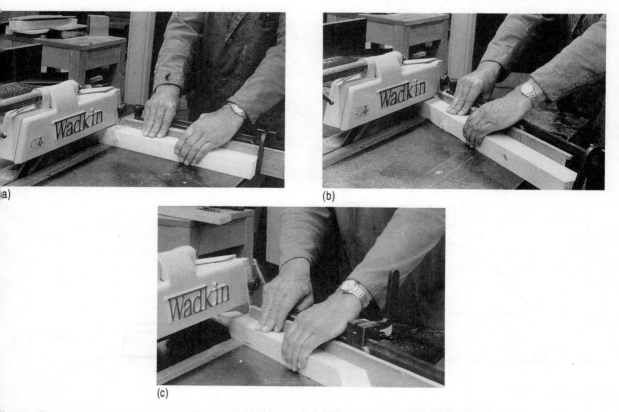

Fig. 3.4 Cross-cutting with a dimension saw, rolling table and fence. *Note* The crown guard must be lowered before cutting begins – in bottom picture it is shown raised only for the sake of clarity

Guarding (Reg. 16)

Figure 3.5 shows how the riving knife should be positioned in relation to the size of the saw blade. Its thickness must exceed that of the blade if a 'parallel-plate' saw blade (Fig. 3.8) is used, so as to keep the saw cut (kerf) open and prevent it closing on to the saw blade. (If the saw cut was allowed to close (with case-hardened timber etc.), the upward motion of the back of the saw blade could lift the sawn material and possibly project it backwards towards the operator.)

During the sawing process, the crown guard must provide adequate cover to the saw teeth. It should extend from the riving knife to just above the surface of the material being cut (it may be necessary to use a crown-guard extension piece) – the gap should be as narrow as practicable, but never more than 12 mm (Fig. 3.5).

Saw blades

Saw-blade size (diameter) in relation to the working speed of the saw spindle is very important, so much so that every machine must display a ' notice stating the minimum saw-blade size which can be fitted (Reg. 17(3)).

Saw blades are designed to suit a particular type of work, and in the case of a saw bench a rim speed (peripheral speed) of about 50 metres per second is considered suitable – lower or higher speeds could cause the blade to become overstressed and result in a dangerous situation. Cross-cut saw blades usually require a higher rim speed than rip-saw blades.

To calculate the rim speed (peripheral speed), we must know the diameter of the saw blade and its spindle speed.

Let us assume a saw-blade diameter of 550 mm and a spindle speed of 1750 rev/min.

i) Find the distance around the rim (i.e. the circumference), using the formula for the circumference of a circle:

$$\text{circumference} = \pi \times \text{diameter}$$

π ('pi') may be taken to be 3.142 or $^{22}/_{7}$.

$$\text{Diameter } (D) = 550 \text{ mm} = 0.55 \text{ m}$$

$\therefore$ distance around rim $= \pi \times D$
$$= 3.142 \times 0.55 \text{ m}$$
$$\approx 1.728 \text{ m}$$

This is the distance travelled by a tooth on the rim in one revolution.

ii) Find the distance travelled every minute by multiplying the distance around the rim by the number of revolutions per minute:

$$1.728 \text{ m} \times 1750 \text{ rev/min} = 3024 \text{ m/min}$$

iii) To find the answer in metres per second (m/s), we must divide by 60:

$$\therefore \quad \text{rim speed} = \frac{3024}{60} \text{m/s} \approx 50 \text{ m/s}$$

The formula may be summarised as

rim speed (m/s) =

$$\frac{\pi \times \text{diameter of blade (mm)} \times \text{spindle speed (rev/min)}}{1000 \times 60}$$

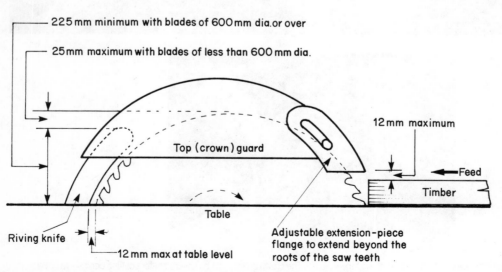

225 mm minimum with blades of 600 mm dia. or over

25 mm maximum with blades of less than 600 mm dia.

Top (crown) guard

12 mm maximum

Feed

Timber

Table

Riving knife

12 mm max at table level

Adjustable extension-piece flange to extend beyond the roots of the saw teeth

Fig. 3.5 Arrangement of riving knife, top (crown) guard, and extension piece (Regs 16(2) and (3))

Choosing the correct saw blade

With regard to its size and the formation of its cutting edge this will depend on one or more of the following:

a) the type of sawing – cross-cutting or ripping;
b) the type of material being cut, i.e. solid timber or manufactured boards etc.;
c) the condition of the material being cut;
d) the finish required;
e) the direction of cut, i.e. across or with the grain.

Tooth shape and pitch (the distance around the circumference between teeth) greatly influence both the sawing operation and the sawn finish. Saw blades are generally divided into two groups: those most suited to cross-cutting and those used for ripping and deeping (both cutting with the grain).

Figure 3.6 shows three different forms of tooth profile used for cross-cutting. Notice that the front face of each tooth is almost in line or sloping forward of the radius line – this is known as 'negative hook' and produces a clean cut.

Figure 3.7 shows how each tooth of a rip-saw blade slopes back from the radius line to produce a true hook shape, which is therefore termed

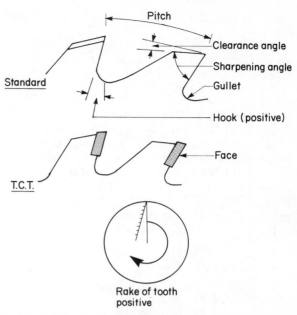

Fig. 3.7 Rip-saw teeth. WARNING A rip-saw blade must never be fitted to a cross-cutting machine because this would lead to a self-feed effect of the blade towards the operator

'positive hook'. It is this shape which enables the tooth to cut with a riving or chopping action. The large gullet helps keep the kerf free of saw dust.

A tungsten-carbide-tipped tooth is ideal for cutting hard or abrasive material – because of its hardness, it stays sharp much longer than the conventional saw tooth. However, its eventual resharpening does involve the use of special equipment and, because of this, the maker of the saw blade or an appointed agent usually undertakes the task.

Knowing the number of teeth and the diameter of the saw blade will enable the tooth pitch to be determined,

i.e. tooth pitch =

$$\frac{\text{rim distance (circumference of saw blade)}}{\text{number of teeth around rim}}$$

where rim distance $= \pi \times$ diameter

For example, if a saw blade has a diameter of 500 mm and 80 teeth,

$$\text{tooth pitch} = \frac{\pi \times \text{diameter}}{\text{number of teeth}}$$

$$= \frac{3.142 \times 500 \text{ mm}}{80}$$

$$= 19.6 \text{ mm}$$

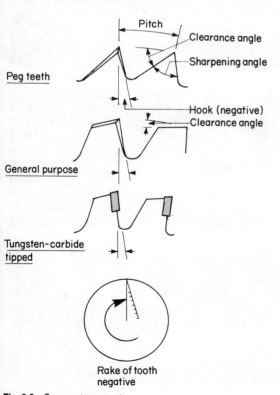

Fig. 3.6 Cross-cut saw teeth

Note   In general, the shorter the pitch the finer the cut.

With the exception of certain saw blades which are designed to divide timber with the least possible waste, the kerf left by the saw cut must be wide enough not to trap the saw blade – Fig. 3.8 shows how this is achieved. The 'hollow-ground' saw blade (Fig. 3.8(a)), which produces a fine finish and is used for cross-cutting, uses its reduced blade thickness, whereas the 'parallel-plate' saw blade (Fig. 3.8(b)) uses its 'spring set' (teeth bent alternately left and right). Tungsten-carbide-tipped (TCT) saw blades (Fig. 3.8(c)) rely upon the tip being slightly wider than the blade plate to provide adequate plate clearance.

Figure 3.8 also shows a saw-blade mounting.

The holes in the saw blade must match the size and location of both the spindle and the driving peg, which is incorporated in the rear flange. The front flange is positioned over the peg before the spindle nut is tightened.

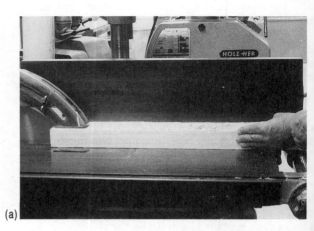

(a)

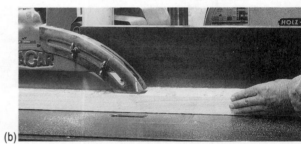

(b)

(c)

(d)

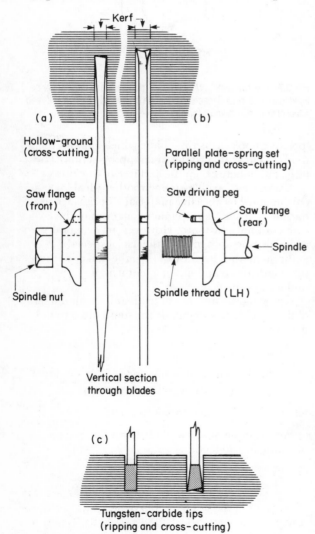

Fig. 3.8   Circular-saw blades and their mounting

Fig. 3.9   Circular-sawing machine in use

Use

Figure 3.9 shows a ripping operation being carried out. In Fig. 3.9(a) the cut is about to be started. As the cut progresses (Figs 3.9(b) and (c)), a push stick (Fig. 3.10) is used to exert pressure on the timber. During the last 300 mm of the cut (Regs 19(1) and (2)) only the push stick is used (Fig. 3.9(d)) – the left hand is moved out of harm's way. Hands are thus kept a safe distance away from the saw blade at *all* times.

By using a jig, jobs which otherwise could not be done safely are often made possible. Figure 3.11 shows a simple jig about to be used – while the cut is being made, a push stick must be used to hold the short length of timber against the jig, as previously mentioned.

Unless the machine is of the movable type which cannot take a blade larger than 450 mm diameter, or is used in conjunction with a travelling (rolling) table, the delivery end of a machine table must be not less than 1200 mm in length from the back of the saw blade (Reg. 20(2)).

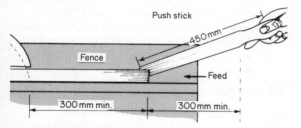

Fig. 3.10 Push stick

Fig. 3.11 Using a jig to cut wedges. *Note* Push stick must be used to hold stock being used for wedges against the jig

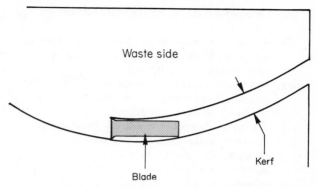

Fig. 3.12 Band-saw blade and kerf (sides of blade must not rub on sides of kerf)

3.3 Narrow band-sawing machines

(See Book 1, Section 5.5, and the Woodworking Machines Regulations Part V.)

For the band-saw to cut safely and effectively, it should be fitted with a blade which is correctly tracked and tensioned and of a size suitable for the work (Table 3.1)

Table 3.1 Minimum radius capable of being cut with a given blade width. (See Fig. 3.12)

Width of saw blade (mm)	3	5	6	10	13	16	19	25
Radius (mm)	3	8	16	37	64	95	138	184

Band-saw blades

Relatively narrow band-saw blades are supplied and stored in a folded coil consisting of three loops. Figure 3.13 and the description which follows briefly explain how this is done.

i) The blade is held firmly, using gloves, with arms outstretched and palms uppermost.
ii) On turning the hands over simultaneously, the blade will twist.
iii) On lowering the blade to the floor it will fold into three looped coils.

Warning Never allow the saw blade to 'slip or turn in or on to your hand, otherwise severe cuts to the skin could result.

Fitting a band-saw blade

Fitting or replacing a saw blade will mean opening the upper and lower guard doors for the purpose of access; therefore, before the start of any such work, the machine must be isolated from its electricity supply – then, and only then, the work may begin:

i)   Open and move aside the top and bottom guard doors.

ii) Remove or move aside other obstructions; for example the blade top guards, guides, and the table mouthpiece.

iii) Using the appropriate handwheel, lower the top wheel the necessary amount to enable the blade to fit on to both wheels.

iv) Fit the blade on to the wheels – making sure the teeth at the table cutting point are facing downwards – then raise the top wheel sufficiently to hold the blade on to the wheels.

v)   Track the saw blade.

vi) Tension the saw blade.

vii) Re-set thrust wheels and guides (see Book 1).

viii) Reposition and secure guards.

(a) Palms uppermost

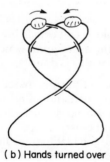

(b) Hands turned over

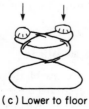

(c) Lower to floor

Fig. 3.13   Folding a band-saw blade with gloved hands

Tracking

This is a means by which the blade is made to run in a straight line between the top and bottom wheels, while passing centrally over their tyres without 'snaking'. The effect of snaking would mean that the back of the blade would intermittently hit the thrust wheel (or thrust roller).

Tracking is checked by turning the top wheel slowly – *by hand* – in a clockwise direction, adjusting the wheel-tilting device until true alignment is achieved. The blade is then tensioned. *Warning*   A blade must never be tracked with the motor running.

Tensioning

Band-saw blades must always be tensioned correctly, otherwise serious damage could result and a dangerous situation could be set up.

The amount of tension applied to the blade via the raising of the top wheel should be such that the blade can be pulled 6 mm away from its running line centrally between the wheels. Alternatively, in the case of machines which have a built-in tensioning scale, tension is applied until the pointer reaches the point on the scale which corresponds with the width of the saw blade being fitted.

Blade length

When measuring the length of a band-saw blade, the amount of tensioning adjustment should be taken into account (Fig. 3.14). The following formula can be used:

$$\text{length of blade required} = \pi D + 2 \times \text{max. distance between wheel centres} - \text{tension allowance}$$

For most industrial saws, a tension allowance of 50 mm is adequate.

For example, to find the length of a band-saw blade when the wheels are 500 mm diameter and the maximum distance between wheel centres (top wheel fully raised) is say 1250 mm,

$$\text{length of blade} = (3.142 \times 500 \text{ mm}) + (2 \times 1250 \text{ mm}) - 50 \text{ mm}$$

$$= 1571 \text{ mm} + 2500 \text{ mm} - 50 \text{ mm}$$

$$= 4021 \text{ mm}$$

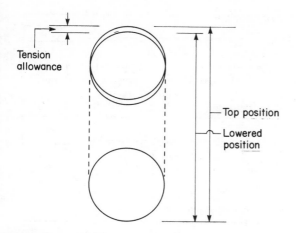

Fig. 3.14 Top-wheel allowance for tensioning purposes

When cutting into a corner (Fig. 3.16), short straight cuts are made first. (This also applies when making curved freehand cuts, Fig. 3.17). An exception to this rule (unless a jig was used) would be when removing waste wood from a haunch (Fig. 3.18), in which case the cuts at A with the grain are made first, to reduce the risk of cutting into the tenon, then the cuts at B. The small portion left at C would, depending on the blade width, probably have to be nibbled away by making a series of short straight cuts.

The operative in Fig. 3.19 is using a band-saw to cut curves freehand.

Use

During all sawing operations the top guard must be set as close to the workpiece as practicable. This not only protects the operative from the saw blade but also provides the blade with maximum support via the guides and the thrust-wheel or thrust-roller assembly (Book 1, Fig. 5.16).

Figure 3.15 shows a straight cut being made with the aid of a ripping fence. Towards the end of the cut, the push stick will be used to push the side nearest the fence.

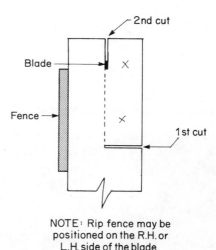

NOTE: Rip fence may be positioned on the R.H. or L.H. side of the blade.

Fig. 3.16 Ripping into a corner

Fig. 3.15 Ripping on a band-saw

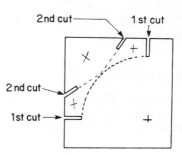

Fig. 3.17 Treatment at the corners of a curve about to be cut freehand

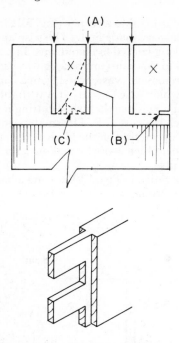

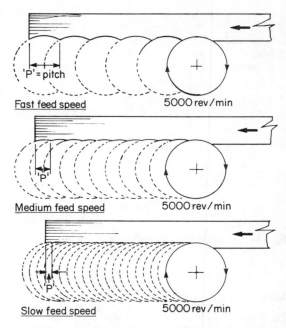

Fig. 3.18   Removing waste wood from a haunched tenon

Fig. 3.20   Example of how a constant cutter speed can produce different surface finishes with different feed speeds

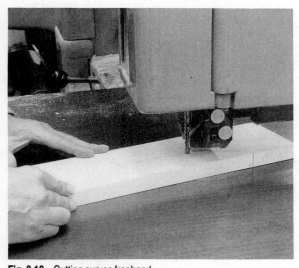

Fig. 3.19   Cutting curves freehand

3.4 Planing machines

(See Book 1, Section 5.3, and the Woodworking Machines Regulations Part VI.)

Planing machines generally fall into three groups:

i)   hand-feed planers and surfacers,
ii)   thicknessers or panel planers,
iii)   combined hand- and power-feed planers.

The quality of finish produced by these machines will to a large extent depend upon the rate at which the timber is passed over the cutters or, in the case of the thicknesser, under the cutters, the speed at which the cutters revolve around their cutting circle (cutting periphery) – on average about 1800 m/min (30 m/s) and the number of blades on the cutter block.

Close inspection of a planed surface will reveal a series of marks in the shape of ripples left by the rotary cutting action of the blade or blades. These marks may have a pitch of 1 to 3 mm. As shown in Fig. 3.20, the shorter the pitch the smoother the surface finish; therefore speeding up the rate of planing the timber would mean degrading its surface finish.

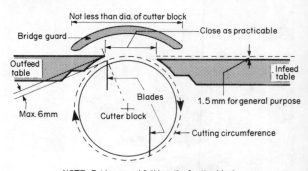

NOTE: Bridge guard full length of cutter block

Fig. 3.21   Cutters, table, and bridge-guard arrangement for surfacing

Guarding the machines

The thicknesser has its cutters enclosed, but the surfacers and combined machines must be provided with a bridge (front and back) guard of a width of not less than the diameter of the cutter block (Fig. 3.21) and long enough to cover the whole of the table gap no matter what position the fence is in or what operation is being carried out (see Regs 26 and 28).

Flatting and edging squared timber

The object of this process is first to produce one perfectly flat face side (flatting) and then, using the fence set at 90° to the table, to produce a face edge at right angles to it (edging).

Before operations of this nature are carried out, table positions and their relation to cutters (see Fig. 3.21 and Reg. 25) should be set to give a cut depth of about 1.5 mm.

Figure 3.22 shows how the bridge guard should be positioned to comply with Reg. 27 when carrying out the various surfacing operations. Figure 3.22(a) shows the position for flatting, Fig. 3.22(b) that for edging, and Fig. 3.22(c) that for when flatting and edging are carried out one after the other.

Hand positions for flatting are shown in Fig. 3.23 – hands must never be positioned over the cutters. Figure 3.23(a) shows the approach position. Once

(a)

(b)

(c)

Fig. 3.23 The process of flatting timber before edging

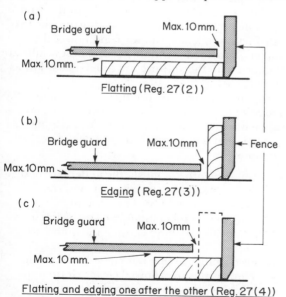

(a)
Bridge guard   Max. 10 mm.

Max. 10 mm.

Flatting (Reg. 27(2))

(b)
Bridge guard   Max.10 mm   ← Fence

Max.10 mm.

Edging (Reg. 27(3))

(c)
Bridge guard   Max. 10 mm

Max. 10 mm.

Flatting and edging one after the other (Reg. 27(4))

NOTE: Maximum gaps must not exceed the distances stated

Fig. 3.22 Positioning the bridge guard for flatting and edging operations (Reg. 27)

the timber has passed under the bridge guard, the left hand is repositioned on the delivery side of the table (Fig. 3.23(b)). As the process continues, the right hand follows (Fig. 3.23(c)) – timber being pressed down on to the table during the whole operation. The whole process is repeated until the desired flatness is obtained.

In Fig. 3.24 edging is being undertaken with the bridge guard set as shown in Fig. 3.22(c) – again, the process is repeated until the edge is both straight and square to its face side.

Fig. 3.24   Edging after flatting

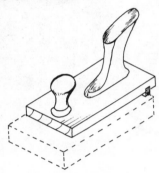

Fig. 3.25   Push block with suitable handhold

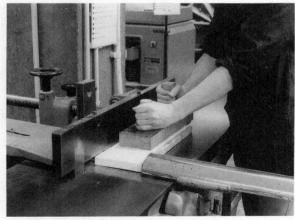

Fig. 3.26   A push block in use

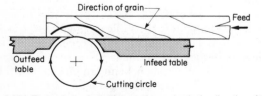

Fig. 3.27   Timber being fed with its grain running in the direction of the cutter

When dealing with slightly 'bowed' or 'sprung' timber (badly distorted timber should be shortened or sawn straight – using a jig if necessary), it should be positioned fully on the infeed table – round side/edge up, or hollow side/edge down – then passed over the cutters by making a series of 'through' passes until straightened.

Where, for reasons of safety, short pieces of timber cannot be planed in accordance with Fig. 3.22(a), a push block (Fig. 3.25) offering a firm and safe handhold should be used. Figure 3.26 shows a push block in use.

Wherever possible, the direction of wood grain should run with the cutters (Fig. 3.27). In this way, tearing of the grain is avoided and a better surface finish is obtained.

Thicknessing

This process involves pushing a piece of timber, face down, into the infeed end of the machine, where it will be met and gripped by the fluted (serrated) infeed roller and mechanically driven under the cutters. Depending on the size and make of machine and the thickness-scale setting, the cutters could remove up to 3 mm off the thickness of the timber – after which it is steadied by the outfeed roller (a smooth-surfaced roller, so as not to bruise the surface of the wood) and delivered from the machine with its upper surface planed smooth and parallel with its underside. Any friction between the underside of the timber and the table bed can be reduced by adjusting the two anti-friction rollers to suit the condition of the timber – provided the timber is dry, best results are obtained with the anti-friction rollers set as low as practicable.

For effective and safe production techniques, the operative should be totally aware of the machine's capabilities (see the manufacturer's operating manual) and how it may be used safely. For example:

a)   For a smooth flat finish, the under face of the timber should be straight, flat, and smooth; feed speed should be slow; and the anti-friction rollers should be set as low as practicable without causing the timber to stick.

b)   Feed timber so that it is cut with the grain.

c)   Timber must be allowed clear exit from the machine – lengths must be limited to well within the distance between the outfeed outboard roller and any obstacle.

d)   Suitable means should be found for supporting long lengths of timber at the delivery end –

either a purpose-made stand or a proprietary roller stand could be used – otherwise the end of the cut will be stepped.

e) Hands must be kept clear of the infeed end of the machine, or they could become trapped between the timber and the table, or even be drawn into the machine. Figure 3.28 shows two of the danger areas.

f) The possibility of a machine allowing timber to be ejected back towards the operator, due to the direction of cutter rotation, is counteracted (Reg. 31) by using either a sectional infeed roller or an anti-kickback device which would allow timber to travel towards the cutters but lock on to any backward movement.

Note Unless they comply with regulation 31(1), older machines built before 24 November 1974 must only be used to plane one piece of timber at any one time, and a notice must be displayed to this effect (Fig. 3.29).

If machines are used to thickness more than one piece at a time, care must be taken to ensure that the position of the pieces on the table is such that they are restrained from being ejected, as shown in Fig. 3.30.

Figure 3.31 shows the motor-drive and feed arrangement to a 'Wadkin' BAO 300 thicknesser. The infeed roller is spirally serrated and its anti-kickback device is in the form of fingers.

Fig. 3.29 Thicknesser displaying a notice stating that 'Not more than one piece of material shall be fed into this machine at any one time' (Reg. 31(2))

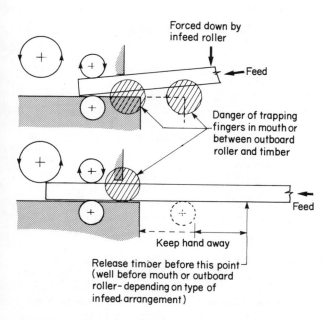

Fig. 3.28 Possible danger areas arising from incorrectly feeding timber into a thicknessing machine

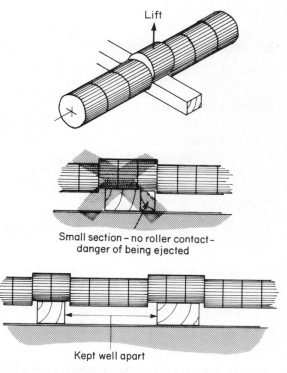

Fig. 3.30 Thicknessing more than one piece of timber at a time

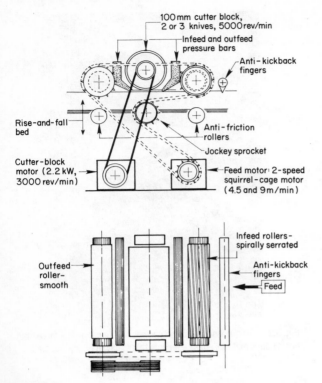

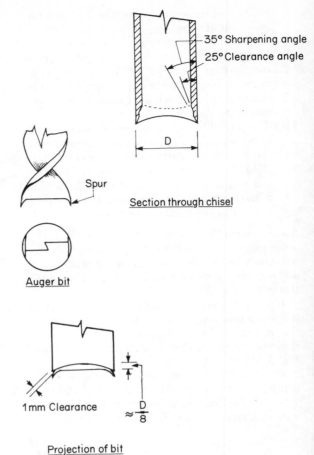

Fig. 3.31   General arrangement of a 'Wadkin' BAO 300 roller-feed planer and thicknesser. *Note*   For clarity, guards are not shown

3.5 Mortising machines

(See Book 1, Section 5.4.)

Chisel mortising

In order to cut the square hole cleanly and safely without damaging the cutters or 'blueing' them (as a result of overheating due to friction), a gap of 1 mm must be left between the spur edges of the auger bit and the inside cutting edges of the square chisel, as shown in Fig. 3.32. This gap can be obtained by positioning both the chisel and the auger bit in the machine together and tightening the auger bit in the chuck when the chisel is 2 mm below its shoulder seating; then lift the chisel to close the gap and tighten the chisel as shown in Fig. 3.33.

Chisel and auger bits should be kept sharp. Chisel sharpening angles (Fig. 3.32) are maintained by using a sharpener like the one shown in Fig. 3.34, which is used with a carpenter's brace – a fine file is used on the inside corners. Bits are sharpened with a fine file in a similar way to a twist bit, as shown in Fig. 2.116 of Book 1.

Fig. 3.32   Auger-bit and square-chisel arrangement

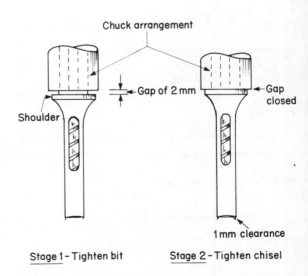

Fig. 3.33   Fitting an auger-bit and square-chisel into a mortising machine

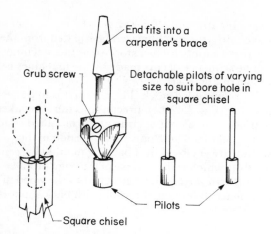

Fig. 3.34 Hollow-chisel sharpener

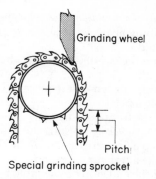

Fig. 3.35 Regrinding a chain

Chain mortising

Chains should be kept sharp, correctly adjusted for tension, and effectively guarded at all times.

Chains are sharpened on the cutting face of each link with an oil-coated slipstone. Regrinding (Fig. 3.35) is done on the grinding assembly at the rear of the machine. When grinding, care must be taken not to cut too deep into the gullet, as this would weaken the chain.

Chain tension between the drive sprocket and the guide-bar wheel should be such as to permit 6 mm of chain slack, as shown in Fig. 3.36.

The chain must be fully guarded at all times during the mortising operation – while the chain guard rests on the workpiece, shrouding the mortise hole, the chaining mechanism is allowed to rise or fall as necessary.

The hardwood chipbreaker should be sited on the workpiece and close to the chain, thus preventing the upward cutting action of the chain from splitting away the end of the mortise hole. Similarly, where through mortise holes are to be cut, it will be necessary to plant and securely fix a length of 32 mm thick packing into or on to the bottom of the table – this will prevent the chain from splitting wood away as it passes through the workpiece. Fixing the packing is of vital importance, otherwise it would be driven out from under the workpiece at dangerously high speed as soon as the chain made contact with it.

Note At no time should the table be traversed while the chain is cutting a mortise, otherwise the guide bar will be strained.

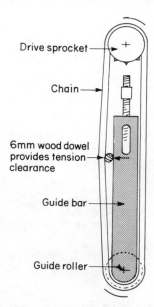

Fig. 3.36 Gauging the amount of chain slack

3.6 Tenoning machines

A single-end tenoner consists basically of a rolling (or sliding) table with fence and clamps, a series of independently driven cutting heads, and a cut-off saw. All these, together with their various means of adjustment, are shown in Fig. 3.37.

Figure 3.38 gives some indication of the types of tenon (in some cases including a scribe) that can be cut on this machine with a single pass between the cutters.

The tenoning process in simple terms will involve the following steps:

  i)   Isolate the machine from its electricity supply.
 ii)   Set the tenoning heads to suit a mortised test-piece (Fig. 3.39).

iii) (*Only if a moulded section is used*) set up the scribing heads with cutters to suit the mould.

iv) Align the cut-off saw to suit the tenon length.

v) Clamp the tenon test-piece tight against the table fence.

vi) Set the front guards.

vii) (*After switching on*) start the machine.

viii) Keep a firm grip as shown in Fig. 3.40 and make a test cut.

ix) Set the end stop on the table to accommodate rail or shoulder lengths.

x) Make successive cuts as necessary.

Note If cutter adjustment has to be made after stage (viii), the machine *must* be isolated from the electricty supply *not* just stopped at the machine.

The sequence of cuts made by the machines is shown in Fig. 3.41.

If pre-moulded stock (Fig. 3.42(a)) is to be tenoned (Fig. 3.42(b)), precautions must be taken to prevent break out of grain at the moulded edge.

The moulded and/or rebated portion of the section are enclosed by a backing mould (Fig. 3.42(c)) which is the reverse of the edge profile. This *Reverse mould* can be attached to the wooden false fence to provide the grain with back support.

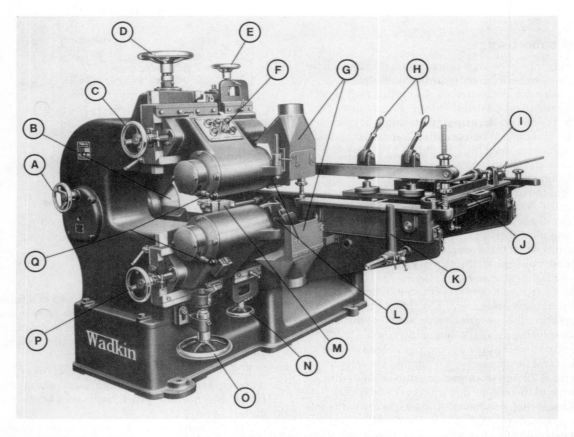

Fig. 3.37 'Wadkin' single-end tenoner   A – handwheel for horizontal adjustment of cutting-off saw   B – cutting-off saw   C – handwheel for horizontal adjustment to top tenoning head   D – handwheel for rise and fall of top tenoning head   E – handwheel for rise and fall of top scribing head   F – independent start and stop push-button control of each motor, with master button to stop all motors   G – guards and exhaust for horizontal cutter blocks   H – quick-acting lever clamps on to sliding table   I – stop bar with turnover stops   J – fence to sliding table   K – adjustable tenon-length stop   L – tenoning heads   M – bottom scribing head   N – handwheel for rise and fall of bottom scribing head   O – handwheel for rise and fall of bottom tenoning head   P – handwheel for horizontal adjustment of bottom tenoning head   Q – brake for tenoning motors

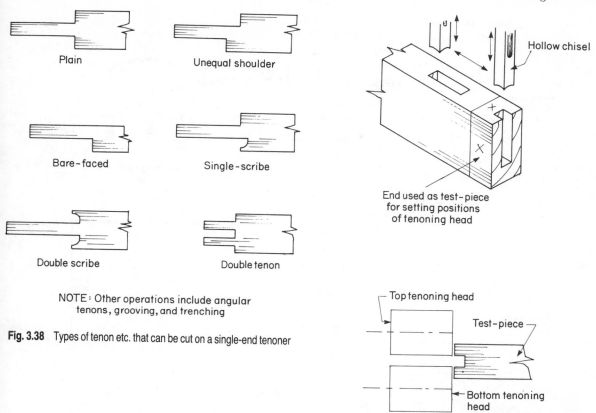

Plain

Unequal shoulder

Bare-faced

Single-scribe

Double scribe

Double tenon

NOTE: Other operations include angular
tenons, grooving, and trenching

Fig. 3.38 Types of tenon etc. that can be cut on a single-end tenoner

Hollow chisel

End used as test-piece
for setting positions
of tenoning head

Top tenoning head

Test-piece

Bottom tenoning
head

Vertical section through tenoning
heads and test-piece

Fig. 3.39 Mortised test-piece for tenoner

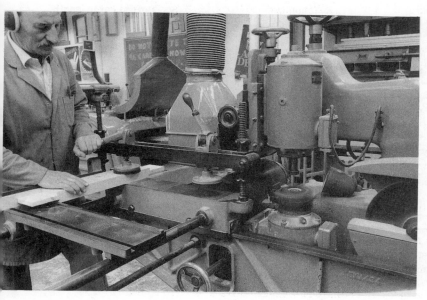

Fig. 3.40 A single-end tenoner in use

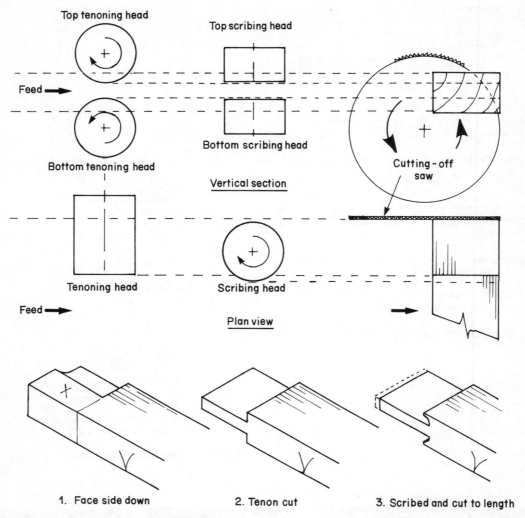

Top tenoning head

Top scribing head

Feed →

Bottom scribing head

Bottom tenoning head

Cutting-off saw

<u>Vertical section</u>

Tenoning head

Scribing head

Feed →

<u>Plan view</u>

1. Face side down

2. Tenon cut

3. Scribed and cut to length

Fig. 3.41 The sequence of cuts made by tenoning machine

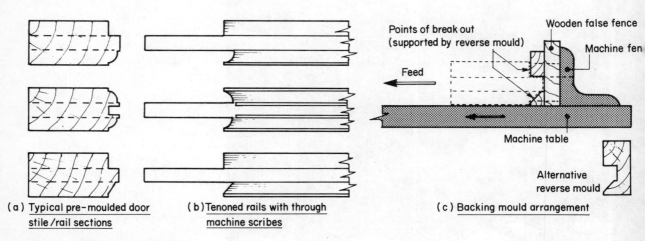

(a) Typical pre-moulded door stile/rail sections

(b)Tenoned rails with through machine scribes

Points of break out (supported by reverse mould)

Wooden false fence

Machine fen

Feed

Machine table

Alternative reverse mould

(c) Backing mould arrangement

Fig. 3.42 Tenoning pre-moulded stock

3.7 Vertical-spindle moulding machines

(See the Woodworking Machines Regulations Part VII.)

Figure 3.43 illustrates the design and control layout of a vertical-spindle moulder. In the main it consists of a machine table through which the spindle protrudes to carry various shaped cutters via a holding device or block (Fig. 3.44).

Spindle speeds of 3000 to 9000 rev/min may be obtained, but only cutter equipment (blocks, cutters, etc.) which has been specifically designed to meet these speeds should be used – to run low-speed equipment at high speed would be very dangerous and could result in the cutters flying off the block. Manufacturers supply cutter equipment to suit a range of spindle speeds; for example, equipment for use only at speeds up to 6000 rev/min, and a separate set for speeds of up to 9000 rev/min.

Both straight and curved moulding, rebating, and grooving can be safely carried out on this machine. Straight work involves the use of a straight split fence, each half of which can be adjusted separately. A circular, or 'ring' fence is used for curved work.

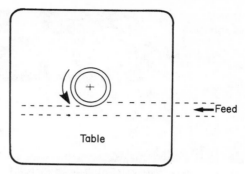

Plan view of spindle rotation

Fig. 3.44 Plan view of the spindle rotation

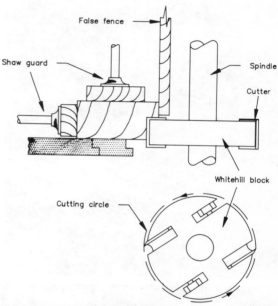

Fig. 3.45 Whitehill cutter block and cutters

Those parts of the spindle, block, and cutters which extend above the working surface of the table must as far as is practicable be fully guarded by using a top guard (bonnet guard), a false fence, a front guard (shaw guard etc.), or a combination of these.

Straight work

Figure 3.46 shows how a false fence closes the gap between the two fence halves yet allows just enough of the cutters' cutting circle to protrude (Reg. 33). Figure 3.47(a) and (b) shows how front protection is provided by a top and side shaw guard forming a tunnel – just enough pressure is exerted by the guards to hold the workpiece steady as it is pushed past the revolving cutters.

Fig. 3.43 'Wadkin' single vertical-spindle moulder   A – push-button controls   B – machine table   C – spindle, block, and cutters   D – topguard   E – combined adjustable fence, guard and hood   F – handwheel for belt tension   G – machine main frame   H – handwheel for rise and fall of spindle   I – lever-operated lock to spindle   J – lock for rise-and-fall motion   K – under-table position of spindle-brake lever (not shown)

Fig. 3.46 A false fence of timber tacked to the main split table to cover the gap

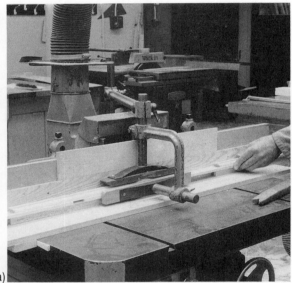

(a)

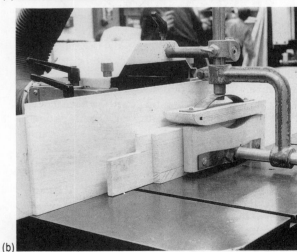

(b)

Fig. 3.47 (a) The use of shaw guards – they are also providing a pressure spring and guide to the work; (b) Shaw guard arrangement for wider stock

Hands are kept clear of the guards by using a push stick (Reg. 38), but, where more than one piece of the same section is to be cut, the pieces which follow may be positioned in turn, one behind the other, until the last piece is to be cut, which will be pushed through with a push stick. Notice the push stick at hand on the machine table in Fig. 3.47(a).

Figure 3.48 shows the use of a power-feed attachment which reduces even further any risk of cutter contact. (The same principle may also be applied to a surface planer or circular-saw bench.)

Where a mould has to be stopped rather than running full length, an arrangement of stops must be firmly attached to the machine (Fig. 3.49). The 'back stop' is very important (Reg. 36) – it prevents the workpiece from being thrown back as the cutters make contact with it. The 'front stop' limits the length of the cut along the workpiece.

Fig. 3.48 A power-feed attachment

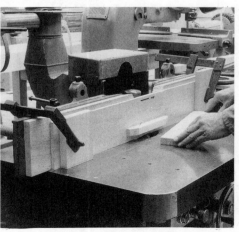

Fig. 3.49 Setting up a fence with back and front stops

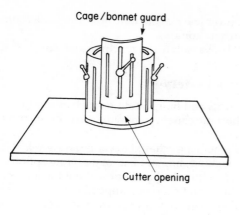

Cage/bonnet guard

Cutter opening

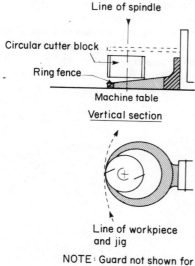

Line of spindle

Circular cutter block

Ring fence

Machine table

Vertical section

Line of workpiece and jig

NOTE: Guard not shown for sake of clarity

Fig. 3.50 Ring fence and guard for curved work

Curved work

Work of this nature will require a jig which holds the workpiece securely while providing safe handholds for the operative. The fence in this case will be in the form of a ring attached to a pillar. Guards must, with the exception of the work face, enclose the spindle and cutting circle by means of a cage, hood, or bonnet guard (Fig. 3.50).

Attachments

By using special purpose-made attachments, work capability can be extended to carry out the following operations:

- **i)** tenoning,
- **ii)** dovetailing,
- **iii)** comb-jointing (a corner locking attachment),
- **iv)** housing of stair strings.

3.8 Safety

No machine must be used unless permission has been granted and the prospective operative is qualified in its use and is fully conversant with and able to act upon all matters of safety.

It is important, therefore, that all trainees make an in-depth study of all aspects of woodworking-machine safety.

Listed below are areas of study together with some of their relevant safety factors.

The Woodworking Machines Regulations 1974 (printed in full at the end of this book)

These must be read and understood in full. If in doubt, ask! Please note that Reg. 44 is obsolete and has been replaced by The Noise at Work Regulations 1989/1990.

Physical condition of the machine and its attachments

Know the location and operation of:

- **a)** the electricity isolating switch,
- **b)** the machine controls,
- **c)** guards and their adjustment.

Setting-up and guarding the machine

- **a)** Know how to isolate the machine.
- **b)** Ensure that the blade/cutters are of the correct type, size, and shape, and are sharp.
- **c)** Ensure that all adjustment levers are locked securely.
- **d)** Check that all guards and safety devices are in place and secure.
- **e)** Ensure that all adjusting tools have been returned to their 'keep'.
- **f)** Make sure that push sticks and blocks are at hand.
- **g)** Ensure that work and floor areas are free from obstruction.
- **h)** Have the machine settings etc. checked by an authorised person.

Suitability of the material to be cut

- **a)** Understand the cutting characteristics of the material.
- **b)** Check length and section limitations.

Machine use

a) Always allow the blade/cutters to reach maximum speed before making a cut.

b) Know the correct stance and posture for the operative.

c) Use an assistant as necessary.

d) Never make guard adjustments until all moving parts are stationary and the machine has been isolated.

e) Never make fence adjustments when a blade/cutter is in motion.

f) Never make fence adjustments within the area around the blade/cutters or other moving parts until these parts are stationary and the machine has been isolated.

g) Concentrate on the job – never become distracted while the machine is in motion.

h) Never allow hands to travel near or over a blade/cutter while it is in motion.

i) Never leave a machine until the blade/cutters are stationary.

j) Always isolate the machine after use.

Personal safety

a) Dress and hair should be such as not to become caught in moving parts or obstruct vision.

b) Finger rings should never be worn in a machine shop, for fear of directing splinters of wood into the hand or crushing the finger if the hand becomes trapped.

c) Footwear should be sound with non-slip soles of adequate thickness and firm uppers to afford good toe protection.

d) Wear eye protection in accordance with the Protection of Eyes Regulations 1974.

e) Wear ear protection in accordance with The Noise at Work Regulations 1989/1990.

4

Site setting-out

Setting-out involves different processes, one or more of which will be necessary before starting any constructional work. They include the following:

a) linear measurement,
b) working to a straight line,
c) setting out angles,
d) setting out concentric curves,
e) establishing a datum,
f) levelling,
g) vertical setting-out.

The accuracy with which these are carried out will determine the final outcome of the work.

4.1 Linear measurement

Measuring distances greater than one metre generally means using a tape. Tapes are available in a variety of lengths, made of steel or fibreglass coated with p.v.c. (Linen tapes are now obsolete – because of their tendency to stretch, they have been superseded by p.v.c.-coated fibreglass types). Tapes over five metres in length have a built-in rewind mechanism.

In use, the accuracy of a tape will depend on

a) the clarity of its graduations;
b) whether it is held in the correct plane (Fig. 4.1);

c) the amount of, or lack of, tension being applied to the tape – very important when measuring long distances;
d) readings being correctly taken from left to right.

Measurement errors frequently occur when measurements are transferred from drawing to site. An example of how errors can be reduced is shown in Fig. 4.2.

Figure 4.2(a) represents a plan view of individually dimensioned wall recesses. If, on transfer, one of these distances was wrongly measured it would not only affect the overall length of the wall but also have a cumulative effect on all those measurements which followed; for example, an error between points B and C would result in points D, E, F, and G being wrongly positioned.

Figure 4.2(b) shows how measurements can be transferred by using 'running measurements' – the tape is run once from A to G (total length), and all intermediate measurements are referred back to A as shown in the build-up of the running total.

Figure 4.3 shows how horizontal measurements over sloping or obstructed ground are carried out. Pegs or posts are positioned on or driven into the ground to act as intermediate measuring stations.

Simple measuring aids like those shown in Fig. 4.4 are very accurate and useful. A small sectioned length of timber can be used to transfer actual

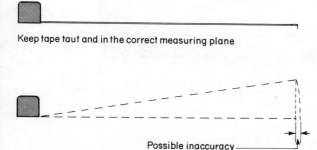

Keep tape taut and in the correct measuring plane

Possible inaccuracy

Fig. 4.1 Using a tape

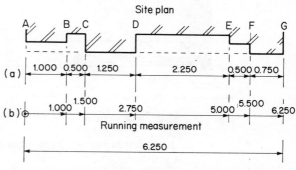

Site plan

Fig. 4.2 Transferring measurements from drawing to site

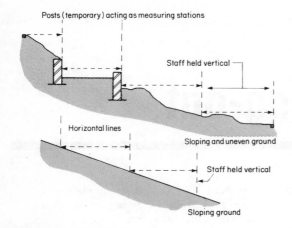

Fig. 4.3 Measuring sloping sites

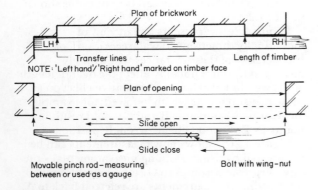

Fig. 4.4 Aids for transferring measurements

measurements – transfers of this nature are often made from site to workshop where an item of joinery has to fit over or into a specific space or opening. A 'pinch rod', on the other hand, consists of two short lengths of timber, one of which slides on the other to enable its total length to be varied. This is ideal for measuring between openings – once set, the pieces are held in that position with nails or a cramp etc.

A more sophisticated method of taking either vertical or horizontal internal measurements is by using a Rabone 'Digi-rod'.

The rod is fabricated from steel, and consists of a body which houses a liquid crystal display, level and plumb vials to ensure accurate readings, and telescopic sections which are manually extended. As the rod is extended the reading is simultaneously registered on the LCD display shown through the window as can be seen in Fig. 4.5.

4.2 Straight lines

Most work carried out by the carpenter and joiner relates to or from a straight line. Straight lines are easily established by using the following methods:

a) a length of string, cord, or wire held taut;
b) a predetermined flat surface such as a straight-edge;
c) visual judgement – sighting.

Figure 4.6 demonstrates how a line is used, and how small obstacles can (particularly over long distances) interfere with the line. A builder's line

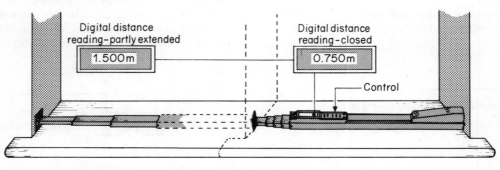

Model range:– 0.750 to 3.200 (illustrated)
1.050 to 5.000 (not illustrated)

Both models can be used to accurately measure vertical and horizontal distances

Fig. 4.5 Taking internal measurements with an electronic digital measuring rod

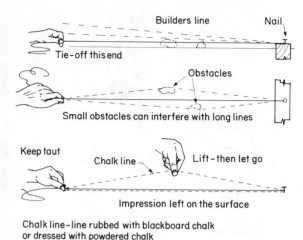

Chalk line – line rubbed with blackboard chalk
or dressed with powdered chalk

Fig. 4.6 Using a builder's line

4.3 Setting out angles

Let us first consider right angles (90° angles), and three simple methods of forming them:

i) by measurement – using the principle of a 3:4:5 ratio;
ii) by using a builder's square;
iii) by using an optical site square.

can be used to mark a chalk line on a floor – the line is coated with chalk (blackboard chalk, or powdered chalk in a proprietary chalking device), stretched tight, then lifted just off the floor midway between its span and let go. On removing the line, it will be found that the chalk has left a clear impression of the line on the surface of the floor.

Straight-edges have been dealt with in Book 1, in connection with ground floors; however, Fig. 4.7 gives a further example of their use – misaligned posts etc. can quickly be recognised.

Neither the line nor the straight-edge is suitable for covering long distances – the line would sag and the straight-edge would be impracticable. Sighting over long distances is possible by a method known as 'boning'. Figure 4.8 shows how boning is carried out by using a minimum of three 'boning rods', usually tee-shaped. The rods are held or stood above the required line. The middle one is positioned at a predetermined point until its adjustment is such that it is in-line with the line of sight – at which point the tops and bottoms of all the boning rods are in a straight line with each other.

Note A straight line is not necessarily a level line.

The 3:4:5 principle

If the lengths of the sides of a triangle are in the ratio 3:4:5 – no matter what units are used – the triangle will be right-angled. This stems from the theorem of Pythagoras which states that, in a right-angled triangle, the square of the hypotenuse (the longest side) is equal to the sum of the squares of the other two sides. Consider the triangle in Fig. 4.9, where side a is 3 units long, side b is 4 units long, and side c (the longest side) is 5 units long:

$$a^2 = 3^2 = 9$$
$$b^2 = 4^2 = 16$$
$$c^2 = 5^2 = 25$$

i.e. $a^2 + b^2 = c^2$

In other words, in a triangle with sides in the ratio 3:4:5, the square of the longest side is equal to the sum of the squares of the other two sides; therefore, such a triangle must be right-angled.

Note The same principle will also work with sides in the ratio 5:12:13.

Figure 4.10 shows how this principle can be applied to a practical situation.

Builder's right-angled square (Fig. 4.11)

This is a purpose-made timber set square. Its size will depend upon where and for what it is used. A useful size could be built around a triangle with sides 900, 1200, and 1500 mm, which fulfils the 3:4:5 rule; for example

side $a = 3 \times 300 = 900$
side $b = 4 \times 300 = 1200$
side $c = 5 \times 300 = 1500$

Its application is shown in Fig. 4.15.

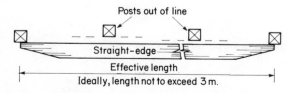

Fig. 4.7 Testing for misalignment with a straight-edge

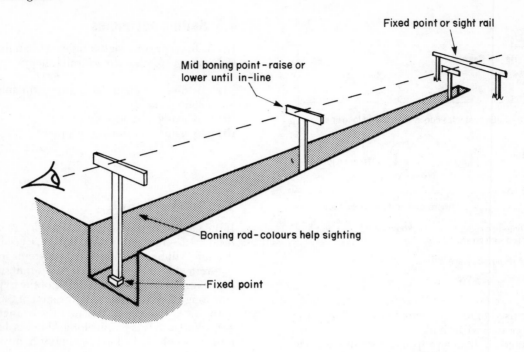

Fixed point or sight rail

Mid boning point – raise or lower until in-line

Boning rod – colours help sighting

Fixed point

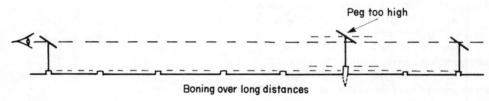

Peg too high

Boning over long distances

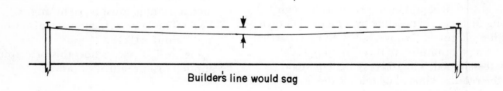

Builder's line would sag

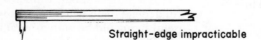

Straight-edge impracticable

NOTE: For visually lining through level or unlevel points over long distances – at, below, or above ground level

Fig. 4.8 Sighting/lining through a straight line over a long distance

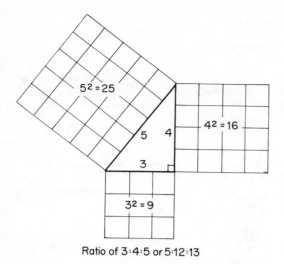

Fig. 4.9 Theorem of Pythagoras

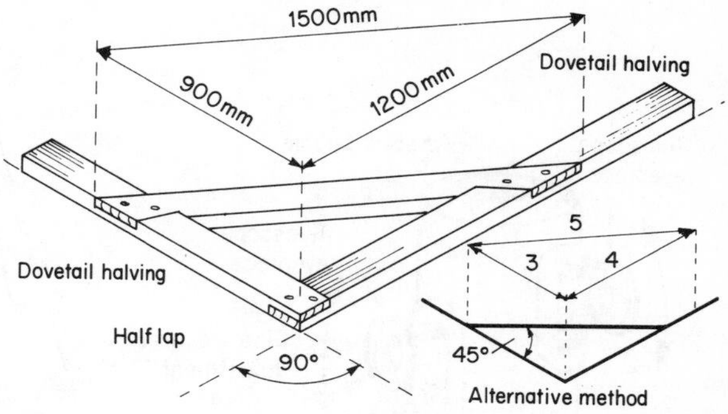

Ratio of 3:4:5 or 5:12:13

Fig. 4.11 The builder's square

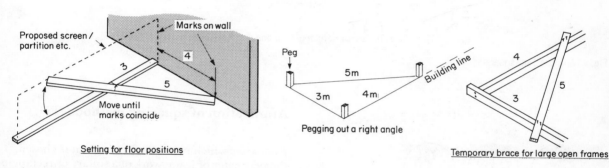

Fig. 4.10 Application of the 3:4:5 principle

Optical site square (Fig. 4.12)

This consists of two fixed-focus telescopes permanently set at right angles to one another, each being capable of independent movement within a limited vertical arc. A circular spirit level in its head enables the datum rod to be positioned plumb over a fixed point or peg.

Figures 4.13 and 4.14 show the stages in using an optical site square to set out a rectangle.

i) Measure the length of one side of the rectangle and mark the ends AB with a 'cross' or 'peg'. A nail should be partly driven into the top of each peg as a sight marker (see insert).

ii) Set up the square at A, positioning the datum rod over the cross or nail. Plumb the rod using the spirit level and by adjusting the tripod legs.

iii) From position A, rotate the head and lock it when the telescope is in line with B. Use the fine-adjustment knob to bring the cross-hairs (viewed through the telescope) exactly over the cross, nail head (see insert), or floor mark – see Fig. 4.14(a).

iv) Keeping the head in the same position, move round 90° to look through the other telescope. Place a mark, or position a peg C, in line with the cross-hairs of the telescope and the correct distance from A – see Fig. 4.14(b). The two sides AB and AC are now at right angles.

v) Repeat the process from B or C, or measure two lines parallel with those right-angled sides formed.

vi) Diagonals must always be checked as shown in Fig. 4.15.

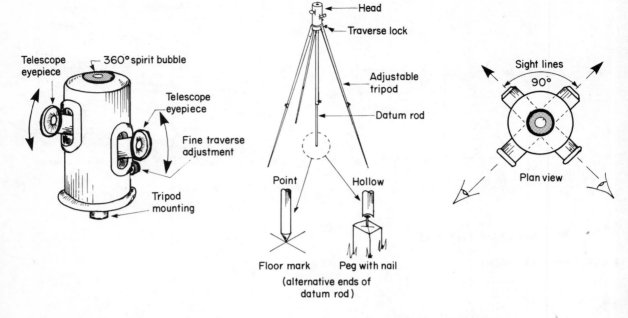

Telescope eyepiece — 360° spirit bubble

Telescope eyepiece

Fine traverse adjustment

Tripod mounting

Head

Traverse lock

Adjustable tripod

Datum rod

Point

Hollow

Floor mark   Peg with nail

(alternative ends of datum rod)

Sight lines
90°

Plan view

NOTE: Telescopes are set at right angles to one another – both move vertically

Fig. 4.12   Optical site square

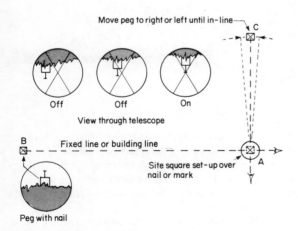

Move peg to right or left until in-line

Off   Off   On

View through telescope

Fixed line or building line

Site square set-up over nail or mark

Peg with nail

Fig. 4.13   Site-square application

As an added bonus, this instrument can also be used to set work plumb (vertical), by raising or lowering the telescope about a horizontal axis so that its line of sight will follow a vertical path.
Note   Figures 4.14(a) and (b) show a site square being used with an offset or extension arm which has allowed the tripod to be positioned away from the datum rod and corner.

Application of squaring methods (Fig. 4.15)

No matter which method of setting-out is chosen, the perimeter or framework of a square or rectangle must have both opposite sides parallel and diagonals of equal length, as shown in Fig. 4.16(a).

Where corner squaring pegs interfere with constructional work, lines are extended outside the perimeter and repositioned on nails or saw kerfs cut into the top of pegs or boards as shown in Fig. 4.16(b). The corners are now located where the lines cross each other, and the lines may now be removed or replaced at will.

Figure 4.17 shows how a rectangular building plot can be set out from a known 'Building-line'.

A 'Building-line' is a hypothetical line, usually set down by the Local Authority to determine the frontage of the proposed building. It may relate to existing buildings or a fixed distance from a roadway.

Once the building-line is established via pegs 1 and 2, the following procedures can be followed:

i)   Frontage distance can be determined with pegs 3 and 4.
ii) Left hand side line is set square from the building-line peg 3 to peg 5.

Fig. 4.14 Site-square in use

iii) Right hand side line is set square from the building-line peg 4 to peg 6.

iv) Depth of building is marked with pegs 7 and 8.

v) Check for square as previously stated in Fig. 4.16(a).

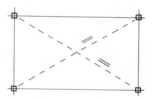

Diagonals same length – frame not square

(a) Opposite sides AA and BB must be parallel and corner–to–corner diagonal same length

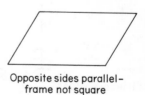

Opposite sides parallel – frame not square

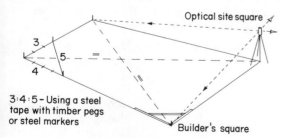

3:4:5 – Using a steel tape with timber pegs or steel markers

NOTE: Diagonal check after all sides have been checked for parallelism

Fig. 4.15 Squaring techniques

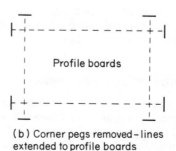

Profile boards

(b) Corner pegs removed – lines extended to profile boards

Fig. 4.16 Setting out a rectangular area

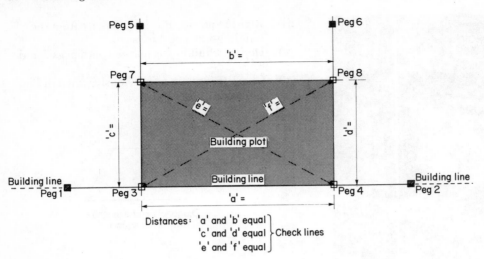

Fig. 4.17 Setting out a rectangular building plot

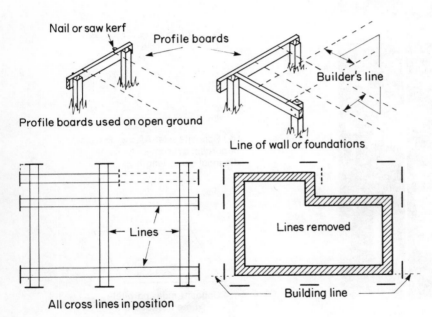

Fig. 4.18 Using profile boards to line out the position of a wall or foundation

Profile boards

These indicate, with the aid of lines, where corners, junctions or changes of direction occur. Figure 4.18 shows how these boards are made and positioned in line with walls or foundations etc.

Angles other than right angles

These are set out by using either angle templates or on-site geometry. Figure 4.19 shows a builder's equilateral triangle and its application. Figure 4.20 shows examples of where and how angles are applied to simple regular and semi-regular polygon shapes.

Regular polygons are figures of more than four sides all of which are equal in length. Each polygon has a specific name in relation to its number of sides, for example:

Number of sides	Name
5	Pentagon
6	Hexagon
7	Heptagon
8	Octagon
9	Nonagon
10	Decagon

Figure 4.20 shows how five-, six-, and eight-sided figures are set out, together with a template suitable for a semi-hexagon.

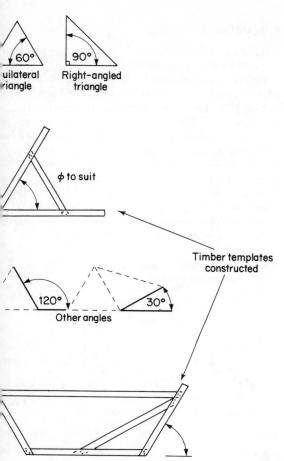

60°
uilateral
riangle

90°
Right-angled
triangle

φ to suit

120°
Other angles

30°

Timber templates
constructed

φ to suit

4.19 Angle templates

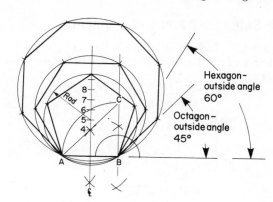

Hexagon-
outside angle
60°

Octagon-
outside angle
45°

Fig. 4.20(a) Regular and semi polygons – geometry

i)   Draw side AB.
ii) Bisect AB to produce the centre line.
iii) Erect the perpendicular BC.
iv) Using radius AB, scribe arc AC across the centre line to produce point 6.
v)   Join AC to produce point 4.
vi) Point 5 is midway between points 4 and 6.
vii) Points 4,5,6,7,8, etc are the same distance apart.

The numbered circle centre points are used to circumscribe the equivalent number of sides AB of the polygon.

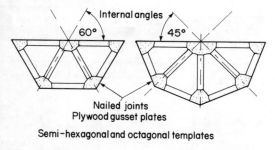

Internal angles

60°   45°

Nailed joints
Plywood gusset plates

Semi-hexagonal and octagonal templates

4.20(b) Regular and semi polygons – examples of use

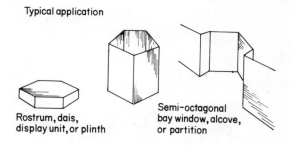

Typical application

Rostrum, dais,
display unit, or plinth

Semi-octagonal
bay window, alcove,
or partition

4.4 Setting-out arcs and segments

Figure 4.21 shows how large circles, semicircles, and arcs can be formed by using either a trammel or a template.

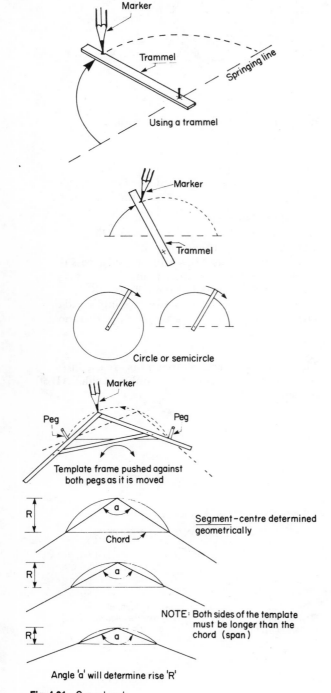

Fig. 4.21 Curved work

4.5 Datums

A datum is a fixed point or horizontal line to whic a height or depth can be referred.

Before any levels can be taken on a building sit a fixed datum point must be established as show in Fig. 4.22. This point is known as a temporary bench mark (TBM) and it must relate to a true Ordnance Survey bench mark (BM) which can b identified by an arrow pointing up to a line whos height above 'mean' sea level at Newlyn in Cornwall is recorded on Ordnance Survey maps. Ordnance Survey bench marks are found cut int rocks or the walls of buildings.

Datum lines shown in Fig. 4.23 need only relat to the work in hand. These are temporary horizontal lines which have been struck at a convenient height so that floor and/or ceiling slopes can be measured and their dimensions be recorded.

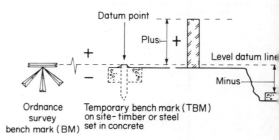

Fig. 4.22 Bench marks

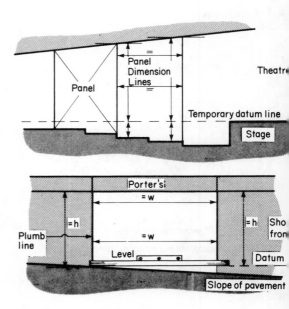

Fig. 4.23 Datum lines

6 Levelling

velling is the act of producing a line or surface
ich is horizontal (level). Any one of the
lowing aids can be used to establish a level line:

    a spirit-level,
    a water level,
    the 'Cowley' automatic level.

irit-level (see also page 93)

is consists of a wood or metal body (rule) with
rallel edges into which are inset one or more
rved glass or plastic tubes containing spirit and a
bble of air. The position of the air bubble
licates whether the spirit-level is horizontal (Fig.
'4(a)).
  The accuracy of the level will depend on the
eness of the bubble tube, the level's effective
gth, and the skill of the operative. Figure
'4(b) shows how, with the aid of a straight-edge,
efficiency can be increased for lengths up to 3 m.
eater distances can be covered by moving both
el and straight-edge together progressively

forward. There is, however, a danger of minor
errors accumulating. This can be avoided if, at
each stage, both level and straight-edge are turned
through 180° as shown (Fig. 4.24(c)) – A–B then
B–A then A–B and so on.

Water level

This works on the principle that 'water always
finds its own level' – which is true when it is
contained in an open system.
  Consider Fig. 4.25(a) – water is contained in a
U-tube and, no matter which way the tube is tilted,
the water level remains in a horizontal plane. In
Fig. 4.25(b) the tube has been divided and base
connections have been made with a flexible hose.
Once again, provided the hose is unobstructed, the
water level will be the same at both ends. This is
the basic principle on which the modern water
level is founded. However, as can be seen from this
model (Fig. 4.25(c)), if one end is raised or lowered
too far, water will spill out – a problem which in

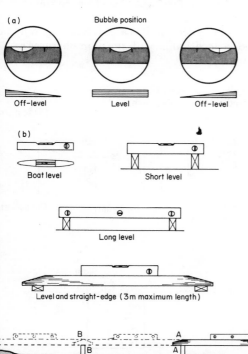

4.24 Levelling with a spirit-level

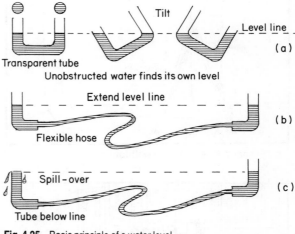

Fig. 4.25 Basic principle of a water level

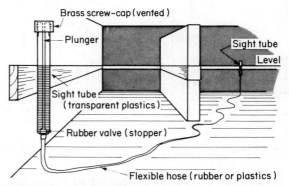

Fig. 4.26 Two-man-operated water level

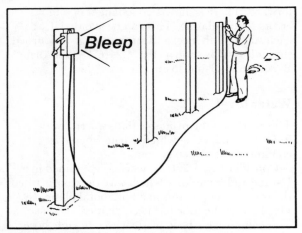

(a) Levelling fence posts

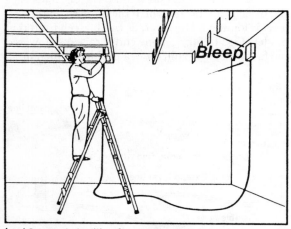

(b) Suspended ceiling from wall datum

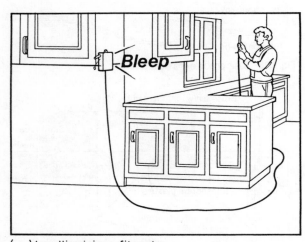

(c) Levelling joinery fitments

Fig. 4.27 'Rabone Chesterman' one-man-operated electronic water level – examples of use

the main has been overcome with the modern water levels.

A modern two-man-operated water level (Fig. 4.26) consists of two transparent (and in some cases graduated) plastic sight-tubes. Each tube h a brass sealable cap with vent holes, and attache to each cap is a brass rod (plunger) with a rubbe stopper which acts as a valve by preventing air entering the hose when the cap is shut down and the level is not in use.

The long rubber or plastic hose – up to 18 m – makes this level particularly useful for levelling floors, ceilings, formwork, etc., especially around corners or obstacles.

Figure 4.27 shows the use of a one-man-operat water level (Electronic Water Level) which has a working distance of up to 18 metres. Once the Electronic Water Level is fixed to a known levelli point (datum) the water filled hose is moved to where the level is to be transferred. When the wa (seen through the transparent tubing) reaches a level that coincides with that of the Electronic Water Level, an audible signal (bleep) is emitted This level can be used in similar situations as the two-man-operated water level – even round corners.

Note   The Electronic Water Level must only be used in accordance with the manufacturer's instructions.

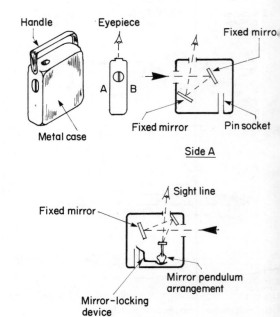

Fig. 4.28   Cowley level and mirror arrangement

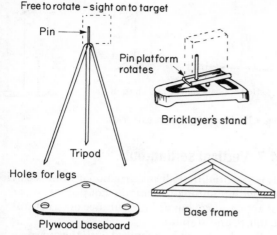

Free to rotate – sight on to target

Pin

Pin platform rotates

Bricklayer's stand

Tripod

Holes for legs

Plywood baseboard

Base frame

Baseboard – frame prevents legs slipping
(NOTE: Slope attachment available)

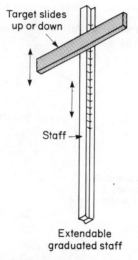

Target slides up or down

Staff

Extendable graduated staff

Fig. 4.29 Cowley-level accessories

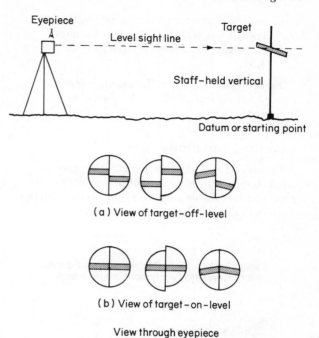

Eyepiece

Level sight line

Target

Staff – held vertical

Datum or starting point

(a) View of target – off-level

(b) View of target – on-level

View through eyepiece

Fig. 4.30 Level application

'Cowley' automatic level

This provides a simple means of levelling distances of up to 50 m with an accuracy of 6 mm in every 30 m. It requires no setting-up or alignment. The levelling mechanism (Fig. 4.28) consists of a dual set of mirrors arranged inside a metal case in such a way that a reflected 'target' as viewed through the eyepiece will appear to be split (Fig. 4.30(a)) unless the top of the target is in line with the 'sight line' – in which case the split images will join together, as shown in Fig. 4.30(b).

The metal case need not be level – variations will automatically be compensated for by the

pendulum movement of one of the mirrors, which comes into operation as soon as the tripod pin is inserted into the base of the metal case.

Note Never carry the level while it is attached to a tripod or stand – the mirror mechanism could be damaged. Once the level is detached from the tripod pin, the mechanism becomes locked and is therefore safe to travel.

Using the level

The procedure for using the Cowley level is as follows:

i) Set the tripod, pin uppermost, on a secure footing – a base board may be necessary on smooth flat surfaces (Fig. 4.29).

ii) Position the level on the tripod and see that it is free to rotate – the levelling mechanism is now unlocked.

iii) Position the staff and target on the 'datum' or starting point. The staff must be held as near to vertical as possible.

iv) Direct the level towards the yellow face of the target.

v) Sight on to the target. An assistant moves the target up or down until the 'on-target' position appears through the eyepiece as shown in Fig. 4.30.

vi) As soon as the 'on-target' position is

achieved, the target is clamped to the staff ready to be transferred to other points requiring the same level as shown in Fig. 4.31.

Figure 4.32 shows that, by taking a reading of the graduated staff, different levels can be measured.

Levelling situations

A Cowley level is ideal for pegging-out ground work; levelling formwork, wallplates, floor joists, etc.; and establishing a datum line on the wall of a building (Fig. 4.31).

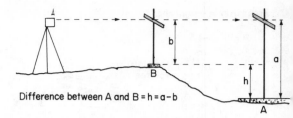

Difference between A and B = h = a − b

Fig. 4.32   Measuring differences in heights

4.7 Vertical setting-out

It is essential that all constructional work is carried out with plumbness in mind – not just for the sake of appearance, but as an assurance of structural balance and stability.

Plumbness can be achieved with the aid of

a)   a suspended plumb-bob,
b)   a plumb-bob and rule,
c)   a spirit level,
d)   a site square (as already mentioned).

The more specialised optical instruments in this field are known as *theodolites*.

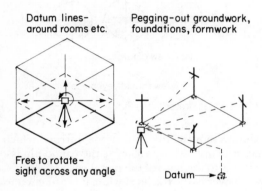

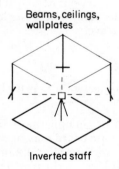

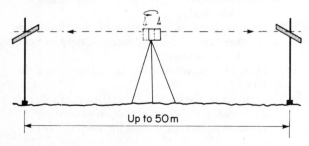

Fig. 4.31   Levelling situations

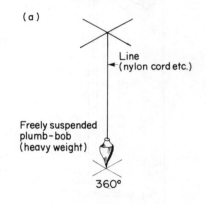

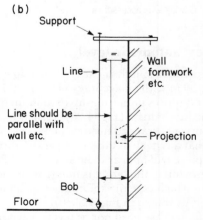

Fig. 4.33   Plumb-bob and line

Plumb-bob

A plumb-bob is a metal weight which, when freely suspended by a cord or wire, produces a true vertical line. It is ideal for indicating vertical drop positions (Fig. 4.33(a)) or as a vertical margin line from which parallels can be measured or referred, such as a wall (Fig. 4.33(b)) or a column box.

Metal bobs can weigh between 56 g and 5 kg or more – choice will largely depend on the type of work and the conditions under which they will be used. A very long plumb line may have to have its bob submerged in a container of water to help reduce the amount of swing.

Plumb-bob and rule

A plumb-bob and rule (Fig. 4.34) allows the plumb line to make contact with the item being tested for plumb. The rule is made from a straight parallel-sided board – a hole is cut through one end to accommodate the bob and allow the line to make contact with the face of the rule, and saw kerfs cut in the top grip the line. A central vertical gauge line is used as the plumb indicator.

Spirit-level

With the exception of the very short level, all spirit-levels can be used to test plumbness (Fig. 4.35). Accuracy will depend greatly on the level's length and bubble setting – some are factory-sealed; others are provided with means of adjustment.

To test a bubble for accuracy, the following steps can be followed:

i) Position the level against a firm straight vertical object. Note the position of the bubble in relation to the central margin on the tube.

ii) Keeping the level in a vertical plane, turn it through 180°. The bubble has now been turned about (end for end).

iii) Reposition the level against the vertical object. The bubble should take up exactly the same position in the tube as before – even the smallest variation is unacceptable.

iv) If adjustment is necessary, the level should be returned to the maker or his agent; in the case of adjustable levels, the maker's instructions should be carefully followed.

Inaccuracies usually only occur as a result of accidental damage or misuse.

Note By substituting the horizontal plane for the vertical plane, the same testing principles can be applied to levelling.

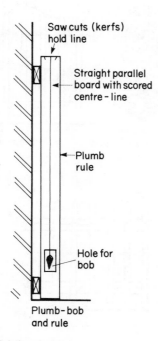

Fig. 4.34 Plumb-bob and rule

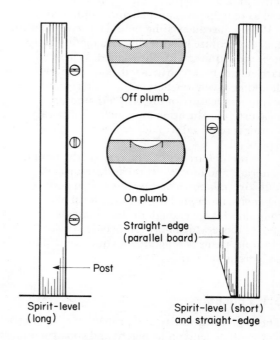

Fig. 4.35 Testing plumbness with a spirit level

5

Fences and hoarding

Land – whether used for building, farming, or leisure – will need to be divided up into partially or fully enclosed areas for the purpose of

a) fixing its boundaries,
b) security of on-site materials and equipment,
c) protection of the general public and the work-force,
d) restricting the movement of animals etc.

Where such areas are to be used as building sites, the extent to which they need to be enclosed often depends on site location. Rural (countryside) sites may require only limited security and protection – although free movement of animals and cattle may cause problems. Urban (town and city) sites, on the other hand, usually require optimum protection to restrain the public from inadvertently wandering into what is often a highly active area and so protect them from accidents – as well as to help prevent theft and vandalism.

If a site has an existing boundary fence or wall, this may require heightening and modifying to meet the requirements mentioned. Where new barriers are to be constructed, consideration should be given to whether they will remain as a permanent feature or are only a temporary measure. The duration for which a temporary barrier is required will depend very much on the type of job.

Perimeter barriers can be very costly, both in materials and labour. Rural and suburban (outskirts of a town or city) sites often permit less expensive inner enclosures in the form of a compound large enough to contain items of value and/or those which may put people at risk – for example, items of machinery, equipment, and dangerous substances, etc.

5.1 Fences

The degree of security and protection offered by a fence will depend on its height and construction. Apart from some recent developments with plastics and methods of preserving wood, styles and constructional methods have changed little over the last twenty years.

Broadly speaking, fencing can be identified as follows:

a) cleft chestnut-pale (Fig. 5.1),
b) chain-link (Fig. 5.2),
c) post-and-rail (Fig. 5.3),
d) palisade and close-boarded (Fig. 5.4),
e) ranch (Fig. 5.5),
f) panel (Fig. 5.6).

Post-and-rail and ranch fencing serve only as visual barriers – their ladder design would do little to discourage would-be intruders. The remaining types can be adapted to form a physical barrier.

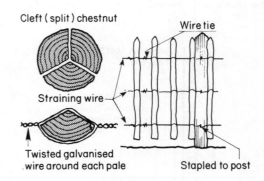

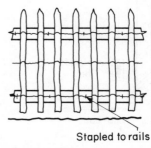

Fig. 5.1 Cleft-chestnut-pale fencing

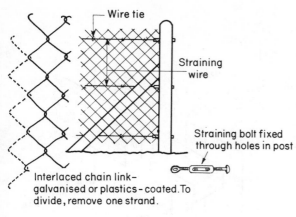

Wire tie

Straining wire

Straining bolt fixed through holes in post

Interlaced chain link-galvanised or plastics-coated. To divide, remove one strand.

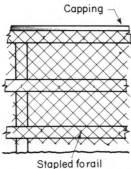

Capping

Stapled fo rail

Fig. 5.2 Chain-link fencing (galvanised or plastics-coated)

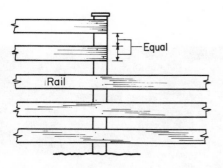

Equal

Rail

Fig. 5.5 Ranch fencing (boards and gaps of equal width)

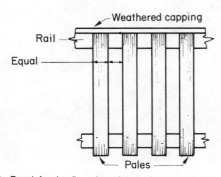

Weathered capping

Rail

Equal

Pales

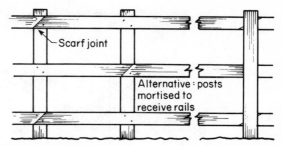

Scarf joint

Alternative: posts mortised to receive rails

Fig. 5.3 Post-and-rail fence

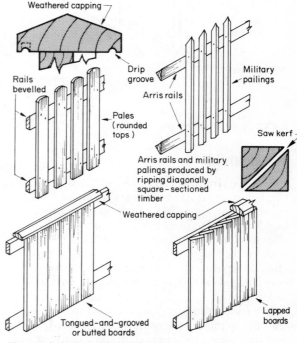

Weathered capping

Drip groove

Rails bevelled

Military pailings

Arris rails

Pales (rounded tops)

Saw kerf

Arris rails and military palings produced by ripping diagonally square-sectioned timber

Weathered capping

Tongued-and-grooved or butted boards

Lapped boards

Fig. 5.4 Palisade and close-boarded fences

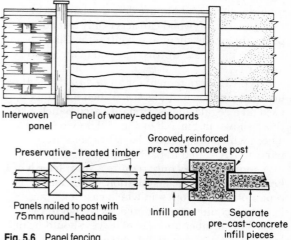

Interwoven panel

Panel of waney-edged boards

Preservative-treated timber

Grooved, reinforced pre-cast concrete post

Panels nailed to post with 75 mm round-head nails

Infill panel

Separate pre-cast-concrete infill pieces

Fig. 5.6 Panel fencing

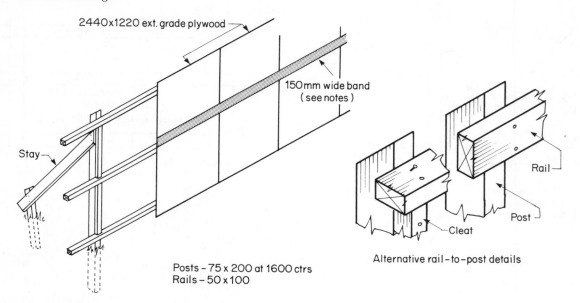

Fig. 5.7   In-situ sheet hoarding

5.2 Hoarding

Under the Highways Act 1980, sections 172(1) and (2), before the start of any work involving the construction, alteration, repair, or demolition of a building situated in a street or court, a close-boarded hoarding or fence should be erected to the satisfaction of the appropriate local authority (LA) to separate the building from any street, court, or public footpath, and thereby ensure public safety.

Hoardings must be designed with structural stability in mind – particularly with regard to resisting wind pressure and, in some cases, crowd pressure. Appearance matters greatly as a public-relations exercise – unsightly second-hand material such as old house doors or badly scarred ex-formwork plywood can serve only as bad publicity for the client and contractor alike – not to mention incurring the disapproval of any local authority.

A typical arrangement for an in-situ sheeted hoarding is shown in Fig. 5.7. When hoarding is used on a regular basis, prefabricated panels like those shown in Fig. 5.8 or Fig. 5.9 may be used. Those which incorporate back brackets are particularly suited to paved areas.

A shopfitter may only require a short length of hoarding during the installation, replacement, or repair of a shop front – in which case a folding hoarding could be considered, as shown in Fig. 5.10. The number and size of leaves should be in keeping with its size, handling, and storage.

Doors and gates in hoardings, to allow personnel and vehicles to enter and leave the site, must not open on to or impede a public footpath or roadway at any time.

Fig. 5.8   Hoarding, viewed from inside a city-centre building site

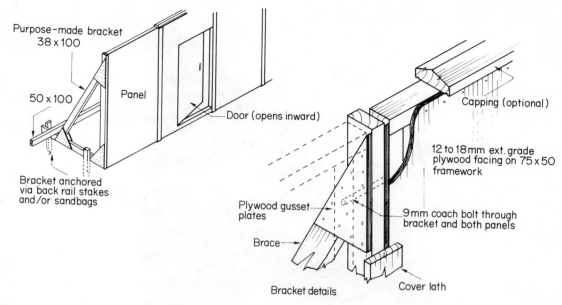

Fig. 5.9 Prefabricated panel-and-bracket hoarding

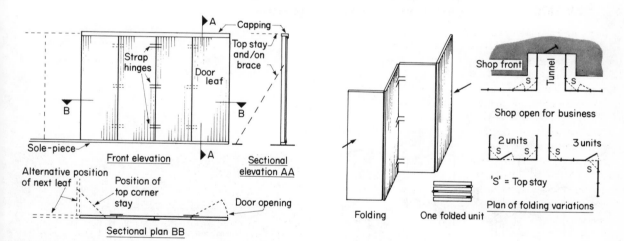

Fig. 5.10 Folding hoarding

Encroachment on to a public footpath

Subject to the width of the existing footpath, and the approval of the local authority, a minimum of 1.2 metres should be allowed for pedestrian traffic.

For reasons of safety, the roadway may have to be encroached upon to provide a walkway, in which case an arrangement similar to that shown in Figs 5.11 and 5.12 may have to be considered and agreed upon with the local authority.

The Highway Authority may ask for some advanced warning signs (works and road narrow signs etc.) for approaching traffic. In some cases the contractor may be asked to provide 'chevron' markings at the start of his temporary walkway.

Provisions must also be made to ensure that both the hoarding and walkways etc. are sufficiently lit during the hours of darkness.

Note   Where temporary structures are sited on, or adjacent to, an unlit pedestrian way, they should be marked with a continuous 150 mm wide white band, with its lower edge running 1.6 m above the ground (Fig. 5.7).

Fig. 5.11   Hoarding around a city-centre development

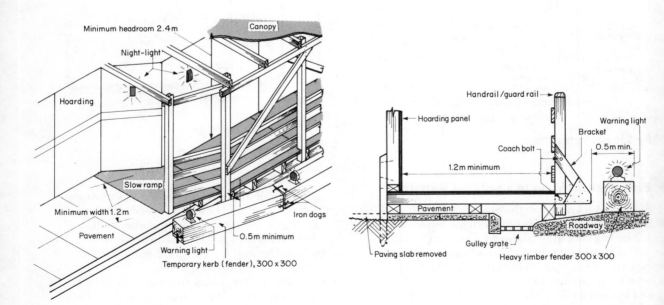

Fig. 5.12   Encroachment on to a public footpath and roadway

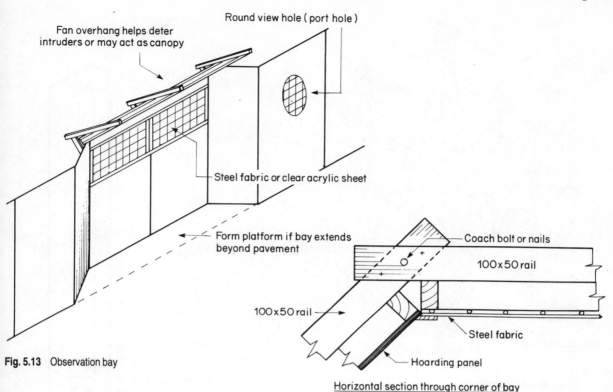

Fig. 5.13 Observation bay

Horizontal section through corner of bay

Observation windows

By virtue of its solid screening, hoarding prevents the general public from viewing site activities; therefore, unless provision can be made elsewhere, it is often advisable to provide openings in the form of windows in the hoarding panels, to discourage the athletic from peering over the top, or peepers from looking through knot holes or panel joints.

If pavement congestion is likely to be a problem, a viewing bay or platform as shown in Fig. 5.13 could be incorporated in the design.

Strategically positioned windows could give an overview of the whole site, making night policing less of a problem.

Fan hoarding

The Construction (General Provisions) Regulations 1961 specify that persons must be protected against falling objects. If, therefore, there is any chance that a person may be put at risk – in this case, outside the confines of the site hoarding as a result of any activity from within – then an arrangement similar to that shown in Fig. 5.14 should be assembled. Its construction must be such as to form a safe catchment to direct whatever may fall back into the confines of the site.

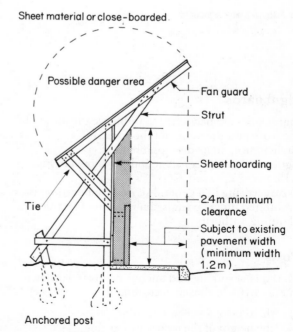

Fig. 5.14 Fan hoarding

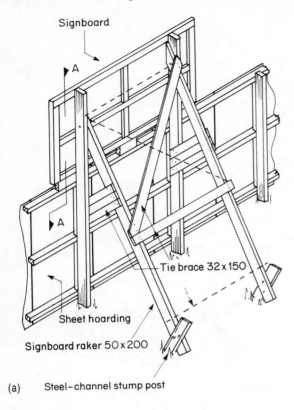

Signboard

A

A

Tie brace 32 × 150

Sheet hoarding

Signboard raker 50 × 200

(a)   Steel–channel stump post

Fig. 5.15   A typical signboard

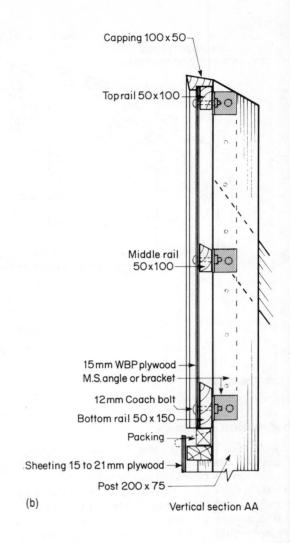

Capping 100 × 50

Top rail 50 × 100

Middle rail 50 × 100

15 mm WBP plywood
M.S. angle or bracket
12 mm Coach bolt
Bottom rail 50 × 150
Packing
Sheeting 15 to 21 mm plywood
Post 200 × 75

(b)

Vertical section AA

Signboards

Signboards are necessary to publicise the nature of the work and the names of the companies participating in its development. A typical signboard and its construction are shown in Fig. 5.15.

Advertising in a general sense on a hoarding is prohibited unless permission is granted by the area planning authority.

Posts

Methods of sinking and anchoring posts as shown in Fig. 5.16 are chosen according to

a) the type of fence or hoarding;
b) the height of the fence or hoarding;
c) the post material – timber, concrete, or steel;

d) the required ground penetration with regard to load and earth condition.

Posts often rely on the added support of a stay or brace to help prevent any deflection and for stiffening.

Sloping ground is not too much of a problem provided a 'stepping' method (Fig. 5.17) is used to offset any gradient – running rails with the slope of the ground should be avoided for the sake of appearance, unless the ground is very steep.

Weathering

Fences, hoardings, and signboards should be designed and constructed in such a way that rain-water is dispersed from their surfaces as quickly as possible – by sloping the tops of posts

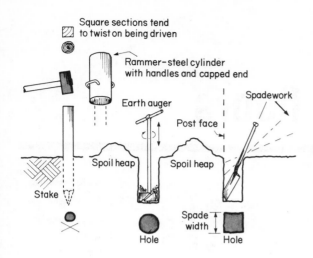

Square sections tend to twist on being driven

Rammer – steel cylinder with handles and capped end

Earth auger

Spadework

Post face

Stake

Spoil heap

Spoil heap

Spade width

Hole

Hole

and rails or by using weathered cappings. End grain is particularly susceptible to decay by wet rot, as shown in Fig. 5.18.

Preservative treatment of timber situated below and just above ground level is essential – see Section 1.7.

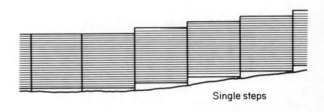

Infill gravel board

Double stepped

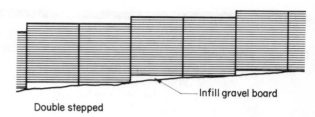

Single steps

Fig. 5.17 Dealing with sloping ground

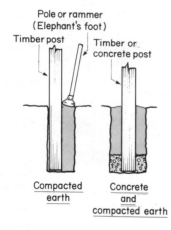

Pole or rammer (Elephant's foot)

Timber post

Timber or concrete post

Compacted earth

Concrete and compacted earth

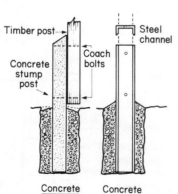

Timber post

Steel channel

Coach bolts

Concrete stump post

Concrete

Concrete

NOTE: If a timber post is to be capable of quick easy removal, it could be temporarily wedged (at ground level) with hardwood wedges into the neck of a sunken (concreted-in) stoneware or plastics sleeve (drain pipe)

Fig. 5.16 Sinking and anchoring posts

Fig. 5.18 The effect of weather on flat-topped timber posts

6

Ground floors

The basic principles of constructing a suspended timber ground floor were dealt with in Book 1.

Figure 6.1 shows a typical layout of a ground-floor plan; Fig. 7.1 shows a corresponding arrangement for a timber upper (intermediate) floor plan.

6.1 Suspended timber ground floor

The floor layout in Fig. 6.1 has made full use of both sleeper and internal walls and has taken advantage of the resulting short spans. As stated in Book 1, perimeter walls can in some cases secure joist ends without a sleeper wall. The under-floor space must be provided with through ventilation from at least two opposing walls – piercing of all sleeper and inner walls is vital if air is to circulate freely throughout the whole underfloor space. Before any flooring (decking) is fixed to the joists, all debris – particularly short ends and wood shavings – must be removed from the under-floor space so as to discourage fungal attack.

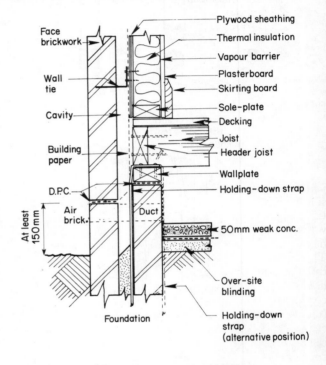

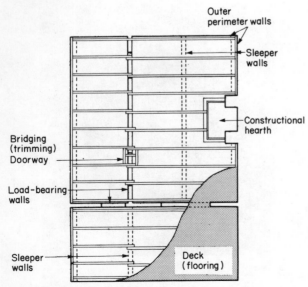

Fig. 6.1 Ground floor joist layout

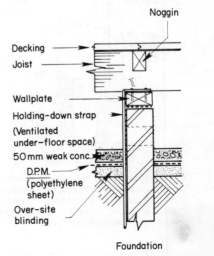

Fig. 6.2 Vertical section through a floor to a timber-framed house

Fig. 6.3 Positioning holding-down straps

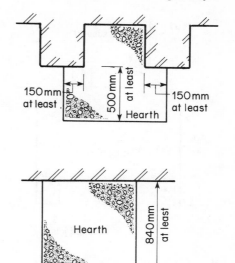

Fig. 6.4 Constructional hearths (solid fuel and oil burning appliances)

Timber-framed houses often use an arrangement similar to that shown in Fig. 6.2: it should be noted that the floor forms part of the main structure – the walls rest upon it. Wallplates and structure must be anchored to the brick or concrete substructure. This is done by using galvanised-steel straps like those shown in Fig. 6.3. In this case sleeper walls have their own foundations, which – provided a continuous oversite damp-proof membrane (d.p.m.) or polythene sheeting is provided – will enable the traditional 100 mm thick oversite concrete to be reduced to a 50 mm layer of weak concrete.

Constructional hearths (Solid fuel and oil burning appliances)

Details of size requirements and the permissible proximity of timber to a fireplace opening etc. are shown in Figs 6.4 and 7.3 – for further details regarding various fuels and gas burning appliances refer to the Building Regulations 1985, approved document J.

Two alternative joist arrangements to meet the above requirements are shown in Fig. 6.5.

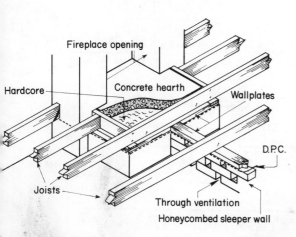

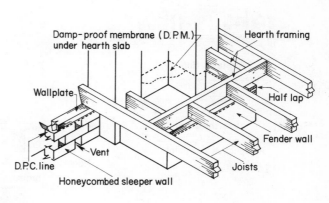

Fig. 6.5 Alternative joist arrangements to a ground-floor constructional hearth

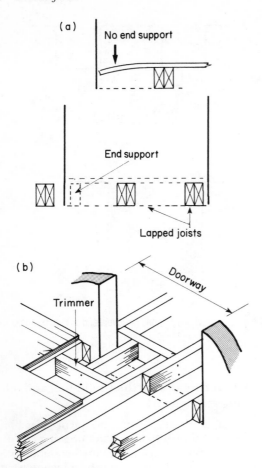

Fig. 6.6   Bridging openings

Doorway openings

Keeping joists to their recommended centres often means that when a doorway is reached it is likely that the edges of sheet decking (flooring) or the ends of floor boards will be without support, as shown in Fig. 6.6(a). It is a simple matter to form a bridge by trimming out with short pieces of floor joist (Fig. 6.6(b)).

Note   For flooring (decking) details, see Section 7.6.

6.2 Split-level ground floors

Floors of this nature are usually constructed for reasons of appearance or as a means of dealing with sloping ground etc. Figure 6.7(a) gives a perspective impression of how a split level may appear.

Construction will depend on which way the joists run and the amount of lift, as indicated in the vertical sections in Fig. 6.7(b). A concrete floor may have to serve as one of the floor levels, in which case ventilation can be provided to the space below the timber via a purpose-made duct or pipe (not less than 100 mm diameter) set in the concrete as shown in Fig. 6.8. Notice how the holding-down straps are bedded into the concrete, and the continuation of the d.p.m. over the d.p.c.

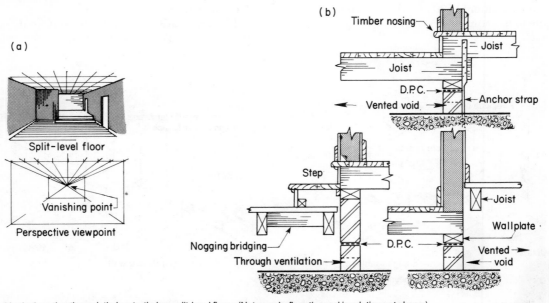

Fig. 6.7   Vertical section through timber-to-timber split-level floors (Note: underfloor thermal insulation not shown)

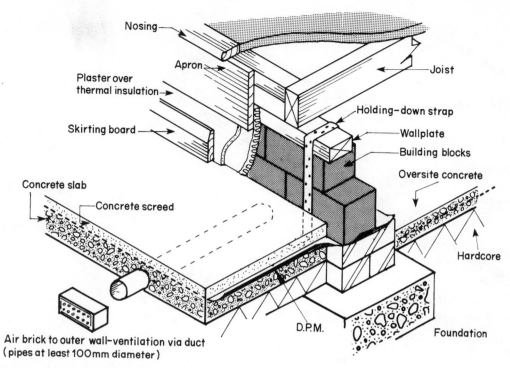

Nosing

Apron

Joist

Plaster over thermal insulation

Holding-down strap

Skirting board

Wallplate

Building blocks

Oversite concrete

Concrete slab

Concrete screed

Hardcore

D.P.M.

Foundation

Air brick to outer wall—ventilation via duct
(pipes at least 100mm diameter)

Fig. 6.8   Concrete-to-timber split-level floor (Note: underfloor thermal insulation not shown)

6.3 Concrete floors incorporating timber

Concrete floors lack the resilience of wood. This problem can be overcome by

a)   attaching a timber floor on to the concrete (Fig. 6.9), or

b)   building a timber floor into the concrete (Fig. 6.10), or

c)   attaching wood blocks to the surface of the concrete.

Figure 6.9 shows how floor joists (bearers) are attached by steel clips, which have been bedded into the concrete before it set. The clips can be flattened (if designed to do so) until required – this reduces the danger of tripping and injuring feet.

In Fig. 6.10, the built-in dovetailed fillets serve as bearers for the flooring (decking).
Note   Preservative treatment is necessary in both cases.

Timber floors built adjacent to concrete floors as shown in Fig. 6.11 can be quite successful provided measures are taken to counteract any movement which may occur at doorways or openings due to moisture or loading (Fig. 6.12) – otherwise floor coverings, carpets, etc. will quickly become worn at these points.

Block floors

Fixing hardwood blocks or strips to a concrete floor is a specialised job. Blocks are usually stuck directly to the concrete surface with a mastic-type adhesive which may act as a d.p.m.

6.4 Thermal insulation

(See also Section 8.8)

Thermal insulation is a means whereby insulation materials – usually in the form of lightweight highly cellular quilts made of fiberglass, mineral wool or granular products such as processed vermiculite (Micafil) etc., are positioned within walls, ground floors and roof, to help prevent warm air escaping from the building during cold weather thereby reducing heating costs. Conversely, thermal insulation materials help to keep the interior of the building cool during warm weather. Figure 6.13 shows how thermal insulation may be introduced into ground floor construction.

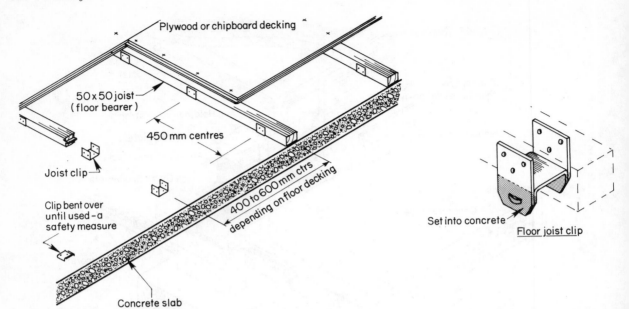

Plywood or chipboard decking

50 x 50 joist
(floor bearer)

450 mm centres

400 to 600 mm ctrs
depending on floor decking

Joist clip

Clip bent over
until used – a
safety measure

Concrete slab

Fig. 6.9   Joists (floor bearers) to a concrete slab

Set into concrete

<u>Floor joist clip</u>

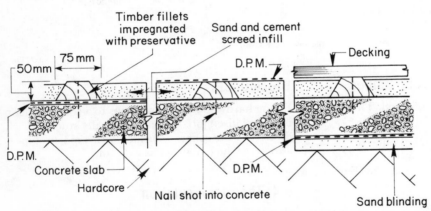

Timber fillets
impregnated
with preservative

Sand and cement
screed infill

Decking

75 mm

50 mm

D.P.M.

D.P.M.

Concrete slab

Hardcore

Nail shot into concrete

D.P.M.

Sand blinding

Fig. 6.10   Vertical section of floors which incorporate timber (Note: underfloor thermal insulation not shown)

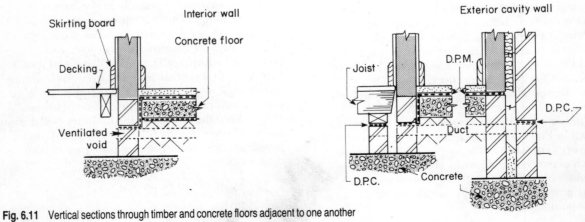

Interior wall

Exterior cavity wall

Skirting board

Concrete floor

Decking

Ventilated
void

Joist

D.P.M.

D.P.C.

Duct

D.P.C.

Concrete

Fig. 6.11   Vertical sections through timber and concrete floors adjacent to one another

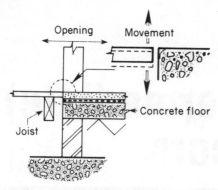

Always avoid butting timber and concrete surfaces together – uneven movement will damage floor coverings.

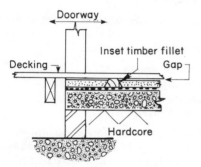

Alternative method – decking allowed to run through the opening (doorway).

Fig. 6.12 Vertical sections to show abutting timber to concrete floors

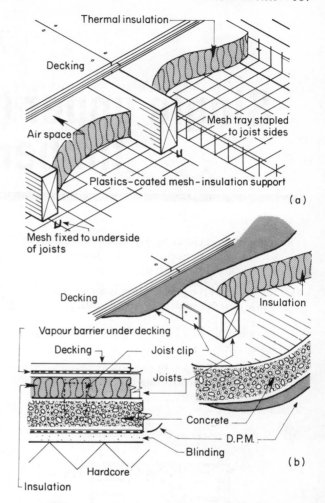

(a)

(b)

NOTE: Method (b) unsuitable in situations where water may be spilt over the decking – water would be held back by the vapour barrier.

Fig. 6.13 Thermal insulation to ground floors

6.5 Access to services

Under-floor space is used to run service pipes and cables, thereby avoiding exposure above the surface and the need for ducts and pipe boxes etc. An example of how service pipes are run under a floor can be seen in Fig. 6.14 (note that water pipes have been insulated).

Wherever services are concealed, provision should be made to allow access for servicing and maintenance; therefore, unless the under-floor space is deep enough to allow a person to work from below, provision should be made from above by way of access traps (Book 1, page 124 and Fig. 7.13). Laying services before the flooring is laid makes it possible to easily identify where traps are required. If, on the other hand, services are laid after the flooring – due to alteration etc. – an insulating layer could make access difficult.

Fig. 6.14 Under-floor pipework

7

Single upper (intermediate) timber floors

The main differences between the layout of a ground floor and an upper floor, as shown in Figs 6.1 and 7.1, are that for an upper floor

a) the depth of the joist section is greater, to allow joists to span a greater distance (no sleeper walls);
b) an opening must be left to accommodate a staircase.

In Fig. 7.1, provision has also been made for the floor to fit around protruding blockwork.

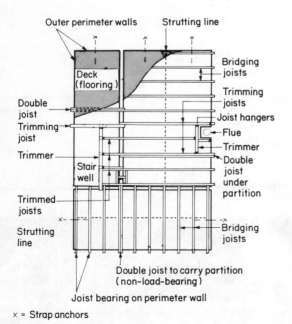

x = Strap anchors

Fig. 7.1 Upper-floor joist layouts

7.1 Floor joists

The determination of a suitable sectional size for an upper-floor joist will, as with all beams, be a matter of calculation, involving

a) the grade of timber,
b) the load supported by the joist,
c) the spacing of joists,
d) the maximum span.

In the Building Regulations 1985, approved document A (1/2), part B, rules are set down to meet particular requirements. With regard to maximum span, it is advisable to keep well within the limits set if undue ceiling deflection (sag) is to be avoided.

Unless a step in the floor or ceiling forms part of the floor design – which is highly unlikely in domestic dwellings – a sectional size of joist selected to span the greatest distance will generally mean that shorter spans will have to use over-sized joists if joist thickness is to remain uniform across the whole upper floor area – which will be so in the majority of cases.

Joist centres should correspond to the dimensions of the decking and ceiling material (plasterboard etc.). Availability and cost are other factors; for example, European-redwood joists may only be available 200 mm wide – if suitably stress-graded and sized at 50 mm × 200 mm, these could span up to about 4 m in a domestic floor. Wider timber is available in spruce and hemlock. Douglas fir is available in sections up to 50 mm × 250 mm but is more expensive.

Note All joists should be regularised in their depth (see Book 1, pages 122–3) to produce a level floor and ceiling.

7.2 Floor support

Joists, positioned 'crown' (curved edge) upwards, are supported either directly by the inner leaf of the structure and/or load-bearing inner walls or indirectly via steel joist hangers.

In some situations it may be advisable to use steel bearing plates between joists and lightweight blocks. Where joists have to be joined over load-bearing walls, they should be lapped and nailed together.

Figure 7.2 shows different methods of supporting the joists, and how the floor framework is used to provide lateral support to walls. Building Regulations require that the whole of the floor framework be attached to the walls. Joists which are built-in should have a minimum bearing of 90 mm. When using joist hangers or when joists run parallel with the wall, they are required to be anchored with steel straps at intervals of not more than 2 m. Nogging should be at least 38 mm thick and over half the depth of the joist deep. For buildings over two storeys in height, see the Building Regulations 1985, approved document A (1/2).

7.3 Trimming

Trimming is an arrangement which enables support to be given to the ends of joists which have been cut short to provide an opening. Stair wells (Fig. 7.1), chimney breasts and constructional hearths (proximity of combustible material in relation to solid fuel and oil burning appliances is shown in Fig. 7.3), ducts, traps, and manholes will all require some form of trimming.

Examples of how these openings may be arranged are shown in Fig. 7.4. All the members in the arrangement have a specific function, namely:

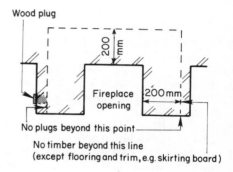

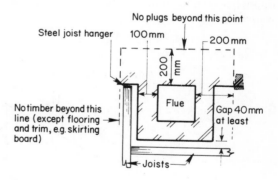

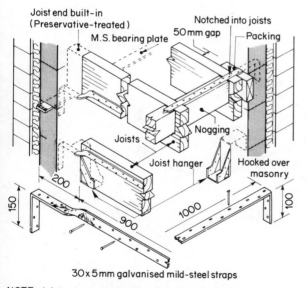

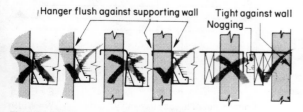

Fig. 7.2 Floor joist bearings and lateral wall support

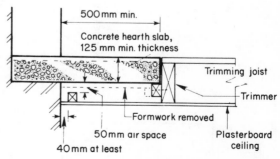

Fig. 7.3 Proximity of combustible material (solid fuel and oil burning appliances)

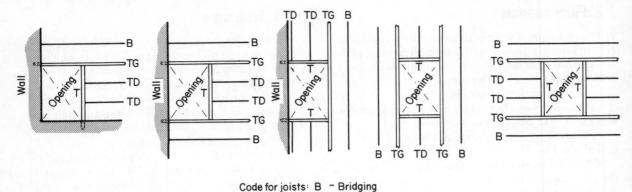

Code for joists: B – Bridging
T – Trimmer
TD – Trimmed
TG – Trimming

Fig. 7.4   Trimming floor openings – chimney breasts, wall protrusions, service traps, etc.

a) *trimmers* (T) are supported by a wall or trimming joist and provide a bearing for trimmed joists,

b) *trimming joists* (TG) are supported by a wall and provide a bearing for trimmers,

c) *trimmed joists* (TD) are supported by trimmers, or by a trimmer and a wall, and provide joist filling between trimming joists or between a wall and a trimming joist.

Trimmers and trimming joists will be more heavily stressed than trimmed or bridging joists, due to load transference between members; therefore both trimmers and trimming joists must either be thicker (usually by 25 mm) than the rest of the joists or be formed from two or more joists nailed together side by side. If the floor is to be heavily loaded, structural calculations will be necessary.

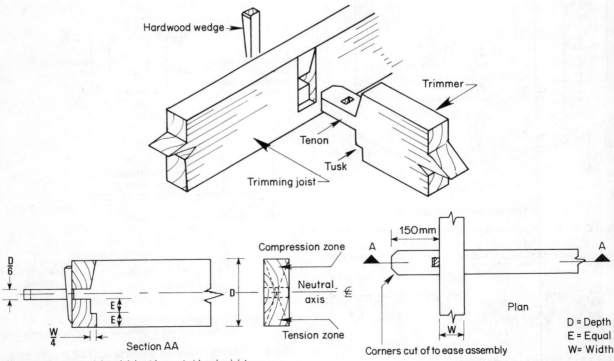

Fig. 7.5   Tusk tenon joint – joining trimmer to trimming joist

Trimming joints

Traditionally the joint between a trimmer and a trimming joist was made with a 'tusk tenon' (Fig. 7.5), and in certain situations this is still used today. However, steel joist hangers have in recent years become very popular, and they are also suited to situations where joists have been doubled-up as shown in Fig. 7.6. Several examples of joining trimmed joists to a trimmer can be seen in Fig. 7.7.

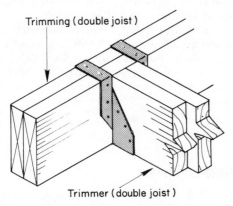

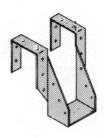

Galvanised steel joist hanger
(timber to timber)

Fig. 7.6 Joist hanger: type of hanger, and method of fixing (size and type of nails/bolts) should be as specified by the designer and/or manufacturer

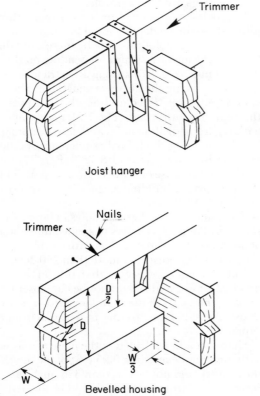

Joist hanger

Stopped housing

Bevelled housing

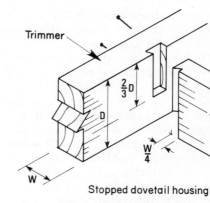

Stopped dovetail housing

Fig. 7.7 Joints – trimmed joist to trimmer

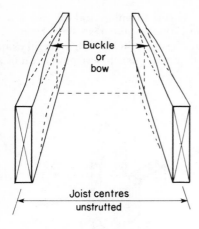

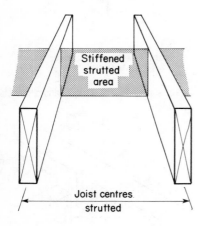

NOTE: Deep thin beams are inclined to buckle under load

Fig. 7.8   Lateral support to joist

7.4 Strutting

Strutting is used to stiffen the whole floor – long, deep, narrow joists will tend to bow and/or buckle unless restrained along their length (Fig. 7.8). Stiffening (strutting) is carried out at intervals according to the joist depth-to-breadth ratio and span (BS 5268: Part 2:1988 'Structural use of timber' Section 2, part 14.8) for example, strutting should be provided at intervals not exceeding 6 times the depth of the joist, when the depth-to-breadth ratio is greater than 5. Thus a 38×200 joist would have a depth-to-breadth ratio of $200/38 = 5.26$. Therefore strutting will be required at intervals not exceeding $6 \times 200 = 1.2$ m.

There are three methods of strutting currently employed:

i)   steel strutting,
ii) solid strutting (solid bridging),
iii) herring-bone strutting.

All three, together with their relevant positions, are shown in Fig. 7.9.

Probably the most effective – although expensive in labour costs – is herring-bone strutting. Figure 7.10 shows how it is cut and fixed.

It should be noted that, no matter which type of strutting is used, the end of the line will need a packing between the last joist and the wall. Care must be taken if folding wedges are used for this purpose, as overtightening could dislodge the blockwork.

7.5 Notching joists

Unlike timber ground floors, where services can be run under the joists, upper-floor services must run either parallel with the joists or, when this is not possible, through them. The only alternative is to run them above the floor on the walls – doorways and openings are then a problem.

Where holes or notches have to be made through joists, the reduced depth should be allowed for in the floor's design, for example by increasing the depth of the joist by 25 mm (depending on the depth of notch) and/or adding an extra joist, thereby reducing the distance apart. Holes for cables should be drilled mid-depth of the joist through the neutral axis (Fig. 7.5). Guide-lines with regard to drilling holes and cutting notches are shown in Fig. 7.11; further reference should be made to British Standard BS 5268 : Part 2 : 1988 'The structural use of timber', Sec., 2, para., 14.9, and the current Building Regulations (Approved document A1/2 part B)

The limits for notches and holes in floor joists shown in Figure 7.11 refers to single family houses of not more than three storeys with simple supported floor joists not more than 250 deep.

Notches: not to be deeper than 0.125 (⅛) of the depth of a joist, and should not be cut closer to the support than 0.07 of the span, nor further away than 0.25 (¼) of the distance of the span from the support.

Holes: should be drilled in the neutral axis with a diameter not exceeding 0.25 (¼) of the depth of the joist, should not be less than 3 diameters (centre to centre) apart and should be located between 0.25 and 0.4 times the span from the support.

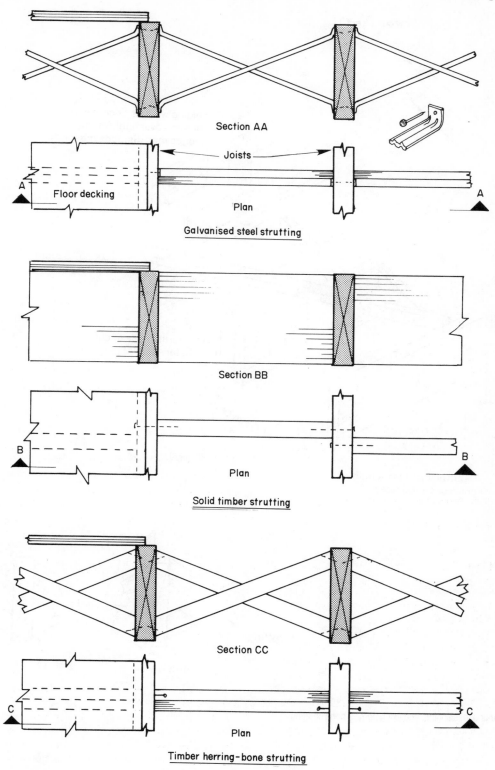

Section AA

Joists

Floor decking

Plan

Galvanised steel strutting

Section BB

Plan

Solid timber strutting

Section CC

Plan

Timber herring-bone strutting

Fig. 7.9 Types of strutting

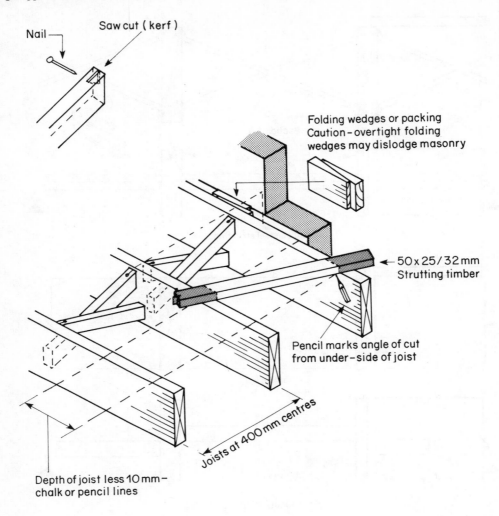

Nail

Saw cut (kerf)

Folding wedges or packing
Caution – overtight folding
wedges may dislodge masonry

50 x 25/32 mm
Strutting timber

Pencil marks angle of cut
from under-side of joist

Joists at 400 mm centres

Depth of joist less 10 mm –
chalk or pencil lines

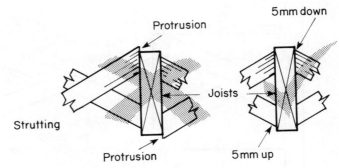

Protrusion

5mm down

Joists

Strutting

Protrusion

5mm up

Fig. 7.10 Marking, positioning, and fixing herring-bone strutting

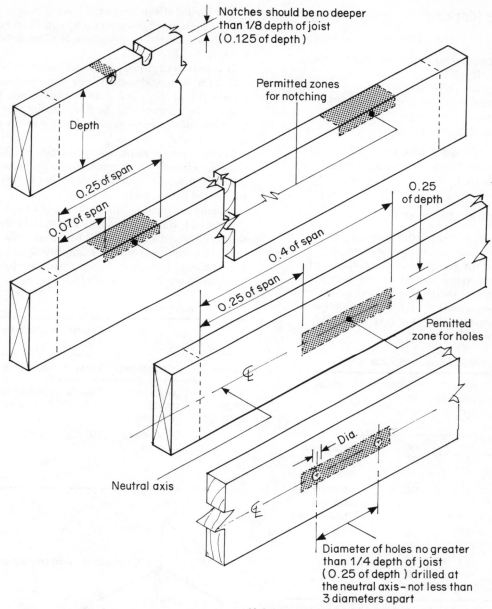

Notches should be no deeper than 1/8 depth of joist (0.125 of depth)

Permitted zones for notching

Depth

0.25 of span

0.07 of span

0.25 of span

0.4 of span

0.25 of depth

Permitted zone for holes

Neutral axis

₵

₵

Dia.

Diameter of holes no greater than 1/4 depth of joist (0.25 of depth) drilled at the neutral axis – not less than 3 diameters apart

Holes must not be within 100mm of a notch

NOTE: The strength and stiffness of the joists must not be put at risk.

This information is a general (rule of thumb) guide only – for single family houses of not more than 3 storeys, and where joists do not exceed 250mm depth

Fig. 7.11   Drilling holes and cutting notches in joists

7.6 Flooring (decking)

As stated in Book 1, materials used as decking are:

a) planed, tongued, and grooved (p.t.g.) floor board (see Book 1);

b) flooring-grade chipboard – tongued-and-grooved or square-edged;

c) flooring-grade plywood – tongued-and-grooved or square-edged.

The processes which follow are common to both ground and upper floors – laying floor board was dealt with in Book 1.

Flooring-grade chipboard

If used in situations where the floor may be subjected to wetting, i.e. bathrooms etc., a moisture-resistant grade should be considered – boards will be marked accordingly on their face. On some contracts moisture-resistant grades are being used throughout the building as a standard procedure. Board thickness in relation to span is shown in Table 7.1.

Table 7.1 Guide to decking-material thickness

Decking material	Finished thickness (mm)	Maximum joist centres* (mm)	Reference
P.t.g. softwood	16 19	450 600	Building Regs, approved document A
Flooring-grade chipboard	18/19 22	450 600	National House Building Council (NHBC)
Flooring-grade chipboard	18/19 22	450 610	BS 5669
Decking plywoods	12.5 16 19–20.5	300 400 600	

* Note   Joists should also be spaced to accommodate the surface dimensions of the decking and ceiling material (plasterboard).

Laying

Tongued-and-grooved boards are laid across the joists as shown in Fig. 7.12 – their ends must meet on a joist. Long edges do not require noggings (horizontal members fixed between joists to support board edges), but the tongues and grooves should be glued together with a PVA adhesive, to help stiffen the floor and prevent joint movement. For nailing details, see Table 7.2.

The laying of square-edged boards is shown in Fig. 7.13 – long edges meet on a joist; short edges must be supported by noggings. Nailing details are given in Table 7.2.

It is important to stagger short joints, to help stiffen the floor and to distribute floor load more evenly. A gap of not less than 10 mm should be left around the perimeter of the floor to allow for moisture movement – gaps will be covered by skirting board.

Where cuts are to be made for traps etc., that portion of board which remains should be not less than 150 mm wide. All free edges should be supported by noggings.

Boards are fixed to joists with nails or screws (Table 7.2), the length of which should be 2½ times the board's thickness. Lost-head nails can be used, but annular-ring nails (improved nails) are recommended because of their resistance to 'popping' (see Book 1, pages 163–7).

Table 7.2 Nailing guide

Material	Edge treatment	Maximum distance between nails (mm)	
		Edges	Intermediate supports
Flooring-grade chipboard	Tongued-and-grooved (all around)	200–300	4 nails to each joist
	Square-edged (all round)	200–300	400–500
Decking plywoods		150	300

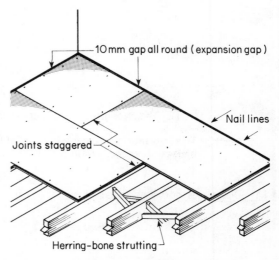

Fig. 7.12 Laying tongued-and-grooved chipboard decking

10 mm gap all round (expansion gap)

Nail lines

Joints staggered

Herring-bone strutting

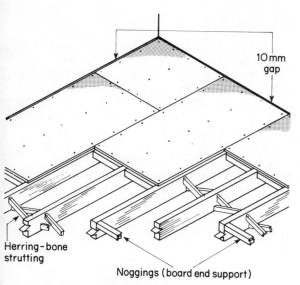

Herring-bone
strutting

Noggings (board end support)

Fig. 7.13 Laying square-edged chipboard decking

Plywood

Like chipboard, flooring grades of plywood are
available with tongued-and-grooved or square
edges. Boards must always be fixed with their face
grain at right angles to the joists. Board ends
should rest on joists, and all edges of square-edged
boards must be supported either by joists or by
noggings. Board joints should be staggered, to
stiffen the floor.

Board thickness is related to span (Table 7.1).
45 mm long 'improved nails' are recommended for
boards 12.5 mm, 16 mm, and 19 mm thick, and
50 mm nails for boards 20.5 mm thick. For spacing
of nails, see Table 7.2.

7.7. Insulation

Thermal insulation will be necessary only where
the floor intervenes between the dwelling and the
external air or a ventilated space; for example, if it
forms part of an overhang or acts as a ceiling to a
porch (Fig. 7.14) it will be required to comply with
the Building Regulations 1985, approved
document L – in which case a similar arrangement

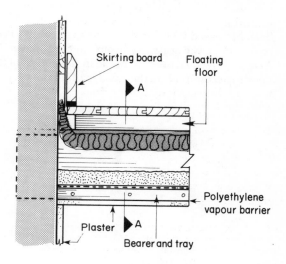

Vertical section

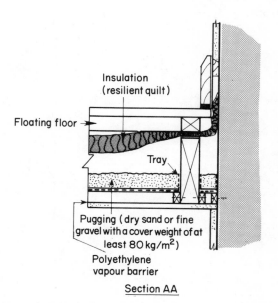

Section AA

NOTE: For further examples, see The Building Regulations
Approved Document E, Section 2

Fig. 7.15 Sound insulation to an upper floor

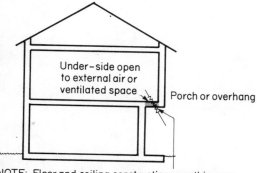

NOTE: Floor and ceiling construction over this area
must comply with approved documents
L (thermal insulation) and B (fire spread)
of the building regulations

Fig. 7.14 Floors requiring thermal insulation

to Fig. 6.14 (with the addition of a suitable ceiling) should be considered.

Sound insulation need only be provided if the floor separates the dwelling from another or as stated in the Building Regulations 1985, approved document E – in which case it must offer resistance to the transmission of both 'airborne' and 'impact' noise. A suitable arrangement is shown in Fig. 7.14, where the decking floats on a resilient insulating quilt draped over the joists to reduce impact noise, e.g. footsteps, and the floor mass has been increased by using dry sand as pugging to reduce airborne noise, e.g. from a record-player. *Note* Allowance must be made in the design of the floor to accommodate the extra loading put on the floor by the weight of the pugging.

7.8 Safety

The height of a habitable room is not generally less than 2.3 m – with the possible exception of the underside of beams – which means that operatives will be working above open joists which are over 2 m above the ground. Therefore, according to the Construction (Working Places) Regulations 1966, where persons are working on or above 'open joisting' through which a person could fall more than 1.981 m, the joisting must be securely boarded over to provide safe access and/or a safe working place.

8

Single timber flat roofs

Although the underside of a 'flat' roof may be flat (horizontal), the upper surface will slope in order to disperse roof water. The amount of fall (slope) should not exceed 10° (otherwise the roof would be classed as a pitched roof) and may depend upon

the type of building and its appearance and location,
the type of decking,
the type of weather-proof covering.

Direction of fall may be affected by the position

joists in relation to decking,
the nearest surface-water drain into which roof water will discharge.

With regard to the laying of roof joists, providing support, anchorage, strutting, and decking, the construction of the roof can generally be likened to that of an intermediate timber upper floor. Figure 8.1 shows diagrammatically how attached and detached flat roofs may appear, with key showing the whereabouts of different edge treatments. Figure 8.2 shows the construction of a flat roof which abuts a house wall.

Roof joists

The Building Regulations 1985, approved document A (1/2), part B, relate joist sectional size grade of timber, loading, span, and centres. Joist centres should be sub-multiples of the dimensions – length and width – of both decking and ceiling material (plasterboard etc.), to avoid unnecessary cuttings.

Fall can be achieved by one or a combination of the following methods:

Sloping joists (Fig. 8.3(a)) – decking and ceiling run parallel. This is suitable for garages, outbuildings, or where there is to be no ceiling or a level ceiling is not required.
Tapered firring pieces (Fig. 8.3(b)) – tapered

pieces of timber (firrings) nailed along the top edge of horizontal joists.
c) *Deepened joists* (Fig. 8.3(c)) – lengths of timber of decreasing section which are positioned at right angles to the fall of the roof and nailed to the top edges of the joists.

If a flat under-ceiling is required, regularised joists (see Book 1, Fig. 7.7(a)) should be used for sloping joists. Regularised joists should also be used where tapered firrings are used, to avoid firring adjustments.

Stiffening of joists can be carried out as for floors. It would be advisable to use packing as opposed to folding wedges at the eaves and verges.

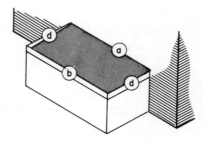

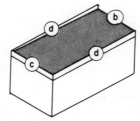

(a) Abutment
(b) Eaves
(c) Heading – as verge or parapet wall
(d) Verge

Fig. 8.1 Key to roof edge details

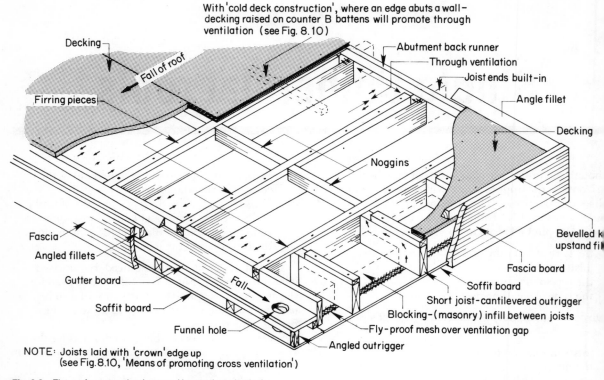

With 'cold deck construction', where an edge abuts a wall – decking raised on counter B battens will promote through ventilation (see Fig. 8.10)

Decking

Fall of roof

Firring pieces

Abutment back runner

Through ventilation

Joist ends built-in

Angle fillet

Decking

Noggins

Bevelled k
upstand fi

Fascia

Angled fillets

Gutter board

Fall

Soffit board

Funnel hole

Fascia board

Soffit board

Short joist-cantilevered outrigger

Blocking-(masonry) infill between joists

Fly-proof mesh over ventilation gap

Angled outrigger

NOTE: Joists laid with 'crown' edge up
(see Fig. 8.10, 'Means of promoting cross ventilation')

Fig. 8.2   Flat-roof construction (exposed isometric projection)

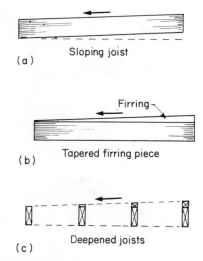

Sloping joist

(a)

Firring

Tapered firring piece

(b)

Deepened joists

(c)

Fig. 8.3   Alternative joist arrangements

8.2 Abutment at walls

Figure 8.1 shows typical abutment locations, for example against a face wall (Fig. 8.1(a)) and a parapet wall (Fig. 8.1(c)). A parapet wall is a wall which protrudes above either the verge or eaves, shown in greater detail in Fig. 8.5, and is terminated by a weathered coping.

Several methods of providing a joist bearing ar shown in Fig. 8.4. They are

a)   *Built-in* (Fig. 8.4(a)) – joist ends should be treated with preservative or wrapped in roofing felt.

b)   *Flange of a steel beam or joist* (Fig. 8.4(b)) – where the roof adjoins an opening in the abutting wall.

c)   *Steel joist hanger* (Fig. 8.4(c)) – built into and *tight* up against the wall face.

d)   *Timber wall piece* (Fig. 8.4(d)) – supported by steel corbel brackets built into the wall. The joist bearing should be reinforced with steel framing anchor plates.

e)   *Steel angle* (Fig. 8.4(e)) – bolted to the wall wi Rawlbolts or similar, depending on the wall material.

f)   *Framed* (Fig. 8.4(f)) – headers bear on steel corbel brackets or angle. The whole roof framework can be constructed at ground leve then hoisted into position – with or without roof decking. This method is suitable for roof or roof sections, of a small surface area.

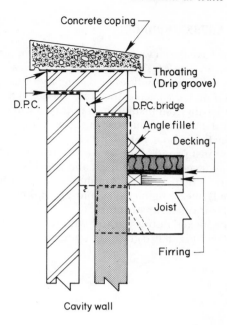

Concrete coping

Throating (Drip groove)

D.P.C.

D.P.C. bridge

Angle fillet

Decking

Joist

Firring

Cavity wall

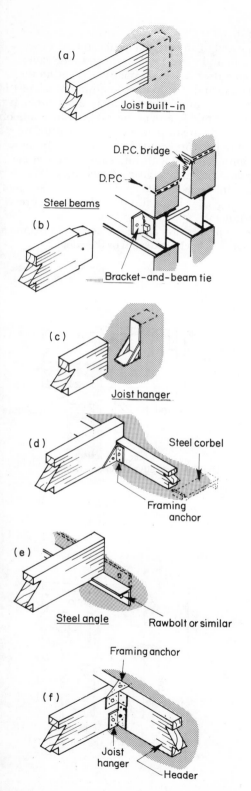

(a) Joist built-in

D.P.C. bridge

D.P.C

Steel beams

(b) Bracket-and-beam tie

(c) Joist hanger

(d) Steel corbel

Framing anchor

(e) Steel angle

Rawbolt or similar

Framing anchor

(f) Joist hanger

Header

8.4 Joist bearings – abutting a wall (firring pieces not shown)

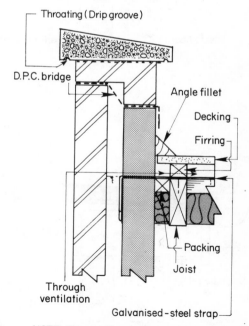

Throating (Drip groove)

D.P.C. bridge

Angle fillet

Decking

Firring

Packing

Joist

Through ventilation

Galvanised-steel strap

NOTE: Thermal insulation to cavity wall not shown. Cross ventilation to cold deck construction not shown

Fig. 8.5 Flat roof abutting a parapet wall (vertical section)

8.3 Eaves treatment

At the eaves, roof water will be discharged into a gutter. Figure 8.6 shows how this area may be treated. Two methods of providing thermal insulation have been shown (see Section 8.8.), but whether thermal insulation is required will depend upon the type and purpose of the building.

The fascia board of flush eaves (Fig. 8.6(a)) stands just proud of the face wall – this will allow for any face-wall unevenness and provide a 'drip'. Figures 8.6(b) and (c) both show closed (although vented) overhanging eaves, but Fig. 8.6(c) shows a timber-framed substructure.

The 'boxed gutter' (gutter with parallel sides) shown in Fig. 8.6(d) forms an integral part of the roof structure – it is expensive to construct compared to an external gutter, but it is very efficient and gives a neater finish to the eaves. Boxed gutters are best constructed within the overhang and should have a minimum fall of 1:6(The gutter outlet (funnel) can be sited either at t ends or at mid length of the gutter, and a balloon galvanised wire or plastics should be fitted over t outlet to help prevent the rain-water fallpipe becoming choked with silt, leaves, or the stone chippings used on some roof coverings. (White stone chippings help keep the roof's surface cool reflecting solar heat, and they also add to fire resistance.)

Note   The position of the surface-water drain ca influence the whole design of the roof.

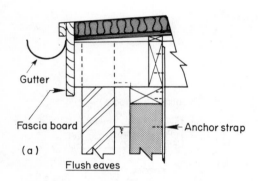

(a)

Flush eaves

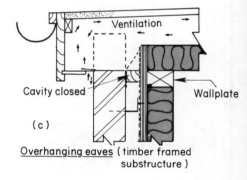

(c)

Overhanging eaves (timber framed substructure)

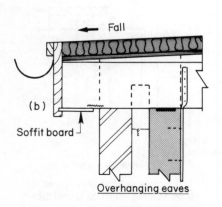

(b)

Overhanging eaves

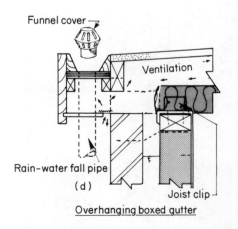

(d)

Overhanging boxed gutter

Fig. 8.6   Alternative eaves details to a thermal-insulated roof (vertical section). Cavity wall insulation not shown

8.4 Verge treatment

A verge is an edge of the roof which lies parallel to the fall of the roof.

Figure 8.7(a) shows a method of providing a kerb edge (an upstand which prevents roof water running off the edge) which abuts the decking material. Joists run at right angles to the fall of the roof.

In Fig. 8.7(b) the kerb is planted on to the decking. The joists now run with the fall of the roof, therefore short joists (outriggers) are used to carry the roof over the end walls and provide a means of fixing the fascia and decking material.

With short roof runs, the fascia can be fixed level (but need not be), in which case the inner upstand is tapered to suit the fall. This is achieved by using a tapered angle fillet or by leaving a tapered margin between the angle fillet and the fascia top. An example of angle-fillet application is shown in Fig. 8.2.

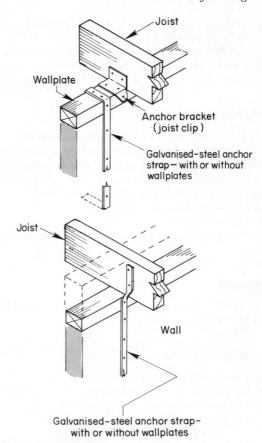

Fig. 8.8   Joist and wallplate anchorage

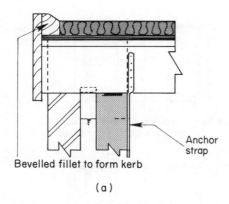

Bevelled fillet to form kerb

(a)

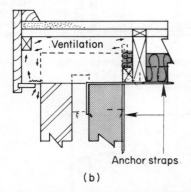

Alternative verge details

NOTE: See Fig. 8.10 re siting roof insulation

Fig. 8.7 Alternative verge details to a thermal-insulated roof (vertical section). Cavity wall insulation not shown

8.5 Roof anchorage

Except when joists are fully built into a wall (Fig. 8.4(a)), this type of roof will only be held down by its own weight, which provides insufficient anchorage to prevent it lifting in strong winds. A suitable anchorage can be achieved by using steel holding-down straps (galvanised), which can also offer lateral support to the walls. Examples are shown in Fig. 8.8. (See also Figs. 7.2 and 9.18 for lateral wall support.)

8.6 Roof decking

Any of the following materials may be specified as a decking.

Tongued-and-grooved floor boards, or square-edged boards

These are laid either with, or diagonal to, the fall of the roof. 'Cupping' of tangential-sawn boards laid across the fall could cause the roof's covering to form hollows of sufficient depth to partially hold back the flow of water off the roof and leave pools.

The methods of laying and fixing are the same as for floors.

Chipboard

Only types recommended for roofing should be used – these possess moisture-repelling qualities. Boards are available which have a covering of bituminous felt bonded to one surface – this gives the board temporary protection against wet weather once it has been laid and the edges and joints have been sealed.

Edge support, laying, and fixing are similar to floors.

Plywood

Only WBP roofing grades should be used, fixed in the same manner as floor decking.

Wood-wool slabs

These are made from cement-coated wood shavings which have been compressed into a slab. Slabs 50 mm thick can span joists at 600 mm centres. They are fixed with special nails.

8.7 Stages of construction

Figure 8.9 illustrates how a flat roof which abuts a wall can be erected. The drawings should be studied in conjunction with Fig. 8.2 which, although thermal insulation has been omitted, shows how through ventilation to the whole roof space can be provided if required.

8.8 Thermal insulation

Unless it serves only as a shelter to a shed, garage, or porch etc., one of the main functions of a roof is to help conserve heat within the structure it covers or, conversely, to offer resistance against heat entering it.

Materials possessing good thermal-insulation qualities are cellular in their make-up and, with the exception of solid wood, are generally of a non- or semi-rigid nature, such as quilts of glass or mineral wool and expanded polystyrene. Rigid slabs are usually in a composite form (cored within a laminated board).

Stage 1 Wallplate (anchored) to front wall piece to back wall – if required

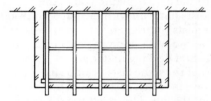

Stage 2. Assemble and strap down joists – centres to suit decking and ceiling boards

Firring pieces to joists

Strutting between joists

Back runner to abutment to carry edge of decking

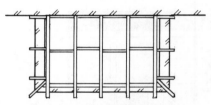

Stage 3(a) Fix outriggers over end walls – form box gutter – if required

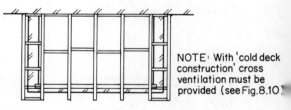

NOTE: With 'cold deck construction' cross ventilation must be provided (see Fig. 8.10)

Stage 3(b) Alternative – fix pre-assembled ladder framework over end walls
Box gutter – if required

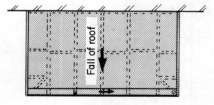

Roof plan

Stage 4 Decking, insulation, soffits, fascia boards

Fig. 8.9 Stages of flat-roof construction

To totally satisfy current (Building Regulations) insulation requirements, the use of timber alone would be both impracticable and uneconomical, therefore lightweight inexpensive materials like those mentioned are introduced within the roof. Where these materials are positioned will depend on whether the roof is of a 'cold-deck', or 'warm-deck' construction.

Cold-deck construction

By placing the insulation material at ceiling level, as shown in Fig. 8.11(a), heat is retained at this point. However, unless a vapour barrier (an impervious membrane, such as polythene sheeting) is positioned at the warm side of the insulation, water vapour will pass through the insulation and, on contact with the cold impervious decking, will turn into droplets of water (condensation). Dampness in this void can mean that there is a danger from fungal attack and that the insulation will become wet and eventually compacted and therefore less effective – not to mention the presence of damp patches on the ceiling below. To prevent the risk of condensation, voids between joists must be thoroughly ventilated. A gap of at least 50 mm should be left between the decking and the thermal insulation to allow for free air space. *Note* 'Through' ventilation can be improved by fixing decking to cross battens set at right angles to the joists (Fig. 8.2).

As shown in Figure 8.10 options include:

i) Fig. 8.10(a) Cross ventilation is provided by leaving gaps equivalent to a 25 mm wide continuous strip along each of two opposite sides. Inner void should be such that a 50 mm gap between the underside of the roof decking and upper side of the insulation is left open to allow free movement of air.

ii) Fig. 8.10(b) Where cross ventilation is not possible (due to an abutting wall), a proprietary head venting system may be installed such as the one shown.

iii) Fig. 8.10(c) An alternative to Fig. 8.10(b), is to use counter battens – these must be strong enough to support the decking and be securely anchored to the sub-structure.

Warm-deck construction

As shown in Fig. 8.11(b), thermal insulation in this case is positioned at decking level, with a vapour barrier to its warm underside. Ceiling voids will not need ventilating. One advantage of this method is that, if access to the roof void is required for services (electric cables etc.), this can be achieved without damaging the vapour barrier.

Note Reference should be made to the Building Regulations with regard to insulation type and relevant thickness for cold-deck and warm-deck construction.

8.9 Roof covering

Layers of bitumen roofing felt stuck to the decking and to each other, with a top layer of bitumen-bedded stone chippings; mastic asphalt; or metals such as aluminium, copper, or lead may be used as a weather-proof covering. Roofing felt can be applied directly to the flat surface of the decking.

With metals there is the problem of thermal movement (expansion and contraction according to environmental temperature change), and provision must be made on the decking to allow movement to take place. This is achieved by the introduction of timber upstands known as 'rolls'. These rolls are positioned at intervals along the fall of the roof Fig. 8.12(a)), the distance apart varying with the type of metal. A roll suitable for lead covering is shown in Fig. 8.12(b) – only one edge of the sheet is fixed. These rolls take care of the longitudinal joint. Because of the thickness of the material – lead, for example – joints across the fall of the roof (width of sheet covering) could form a water check, in which case a step in the decking, known as a 'drip', would be formed as part of the roof's construction as shown in Fig. 8.12(c).

Disposal of water

Falls of not less than 1:60 should be maintained across the whole roof area and the gutters associated with it, otherwise any slight rise in the roof's covering at joints etc. could cause water to pool and become a catchment area for enough silt to encourage and support plant life – which could have a harmful effect on the roof covering and become powerful enough to break open water seals. *Note* See Section 9.9 'Eaves, gutters and down pipes'.

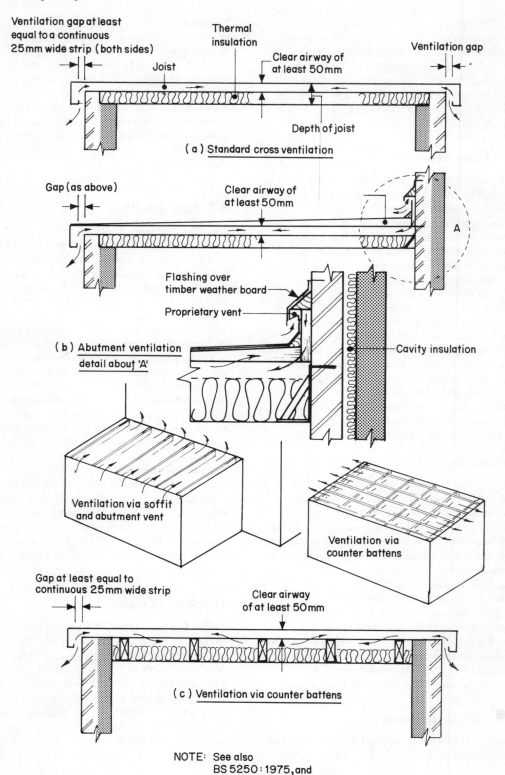

Ventilation gap at least
equal to a continuous
25 mm wide strip (both sides)

Joist

Thermal
insulation

Clear airway of
at least 50 mm

Ventilation gap

Depth of joist

(a) Standard cross ventilation

Gap (as above)

Clear airway of
at least 50 mm

A

Flashing over
timber weather board

Proprietary vent

(b) Abutment ventilation
detail about 'A'

Cavity insulation

Ventilation via soffit
and abutment vent

Ventilation via
counter battens

Gap at least equal to
continuous 25 mm wide strip

Clear airway
of at least 50 mm

(c) Ventilation via counter battens

NOTE: See also
BS 5250 : 1975, and
BS 6229 : 1982

Fig. 8.10 Means of promoting cross ventilation

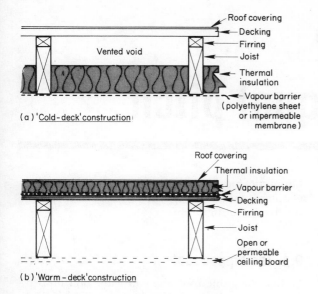

(a) 'Cold - deck' construction

(b) 'Warm – deck' construction

Fig. 8.11   Methods of providing thermal insulation

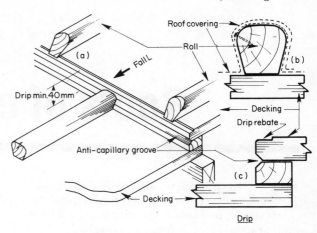

Fig. 8.12   Provisions for thermal movement of a lead-covered roof

9

Roofs of equal pitch

Book 1 dealt with single gable roofs of short span. If the span of such a roof is increased beyond the limits stated, roof failure may occur as a result of either excessive rafter deflection (sag) or substructure spread (Book 1, Fig. 8.3). Rafter deflection can be reduced by one of three ways:

i) increasing the sectional size,
ii) introducing a beam (purlin) midway between the ridge and the wallplate,
iii) substituting a trussed framework (trussed rafter) in place of each adjoining pair of rafters.

The first method could prove expensive and impracticable. The latter two form the basis of this chapter – namely the 'purlin' or 'double' roof and the 'trussed-rafter' roof.

9.1 The hipped roof

Unlike the gable roof, the end walls of a hipped roof finish at the eaves level – above which the roof is sloped to meet up with other inclined roof surfaces.

Figure 9.1 shows a skeletal pictorial view of a gable roof meeting a hipped roof. The intersection of the roof surfaces produces either a 'valley' or a

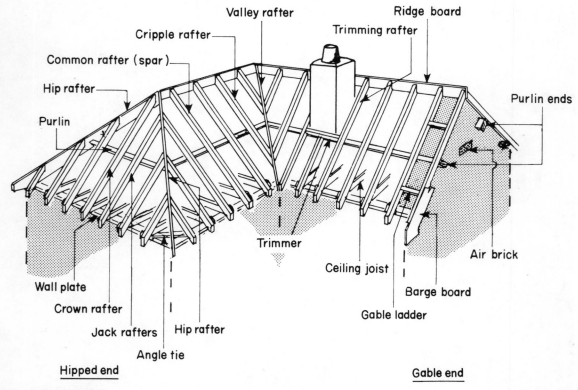

Fig. 9.1 Hip-and-gable roof

'hip'. Hip rafters form a spine to interconnect rafters which have been reduced in length (called 'jack rafters'). Valley rafters are similarly used to connect shortened rafters called 'cripple rafters'. The other members shown – with the exception of the angle tie, which helps prevent wallplate spread – are described in Book 1.

Roof geometry

An understanding of basic roof geometry is needed to ensure that all the members meet up correctly, but before dealing with individual members it is advisable to make a small card or paper model of the developed roof surfaces. This should help to explain how

a) roof surfaces meet up,
b) the plan of a roof can give a false impression of the roof's true size and shape,
c) the true length of a roof's surface and hip can be determined.

To develop the roof surfaces shown in Fig. 9.2, set compasses to the common-rafter length and scribe a semicircle from point P (on the end or front

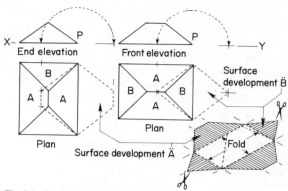

Fig. 9.2 Roof surface development

elevation) to meet the XY line. At the points of intersection, drop perpendicular lines over the plans. The hip-to-ridge intersection points are then extended horizontally until they meet the perpendicular lines, and lines are then drawn to the appropriate corner.

Note If Fig. 9.2 is to be used as a basis for a model, enough room should be left between the elevations and the plan to allow all four surfaces to be developed around the plan, as shown in the pictorial view.

Figure 9.3 shows how the various roof bevels are produced.

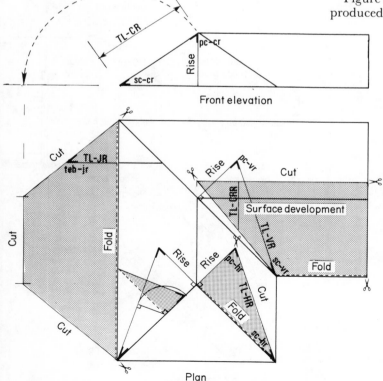

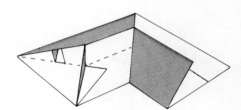

Part model– Bevel–and–cut application

Key:
CR   Common rafter
HR   Hip rafter
JR   Jack rafter
VR   Valley rafter
CRR Cripple rafter
PC   Plumb–cut
SC   Seat–cut
TEB Top–edge bevel
TL   Geometrical true length
□   90° Angle

Fig. 9.3 Roof geometry

To obtain the true length of a hip rafter, draw a 90° line of a length equal to the roof rise from one end of the hip's plan (run) to produce a triangle. The longest side (hypotenuse) of the triangle is then the hip's geometrical true length, and the top and bottom angles are used as hip plumb and seat cuts. Valley-rafter true lengths and bevels are similarly produced.

The hip-rafter backing angle (dihedral angle) is formed as a result of two inclined roof surfaces meeting on a hip. This angle is probably the most difficult of all the angles to produce, and for this reason it has been dealt with in three stages (Fig. 9.4):

i)   Draw a 90° line from A on the true length of the hip line to meet the plan (run) of the hip at B.

ii) Draw line CC through B at right angles to the plan (run) of the hip. From point B with

compasses set to AB, scribe an arc to produce D.

iii) Join point D to C. The resulting angle will be the hip backing angle, as shown by the hip section.

Positioning the hip rafters (Fig. 9.5)

The pair of common rafters which are to receive the hip rafters will have been assembled as an 'A' frame, joined together at the top with a 'saddle board' (plywood or timber) of sufficient depth to allow the plumb cuts of both hip rafters to rest squarely against each other and provide a fixing. It should be noted that, for the hip mitre to meet on the geometrical centre lines, the first pair of common rafters ('A' frame) will have to set back at a distance equal to the saddle-board thickness – otherwise the hip rafters would be too long. The hip edge bevel at this point is shown separately.

The corner wallplate is cut back as shown to

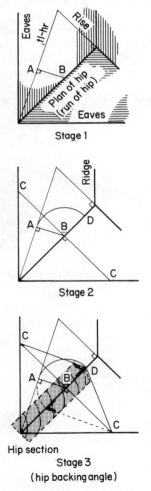

Fig. 9.4   Hip-rafter backing (half dihedral) angle

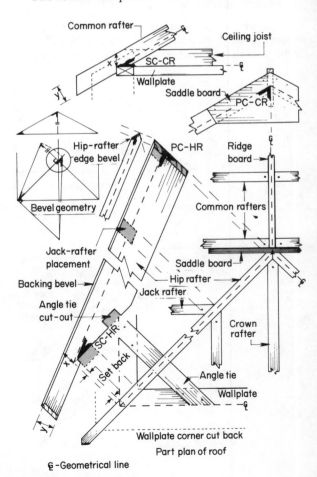

Fig. 9.5   Hip-rafter application

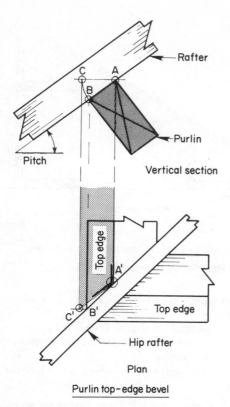

Vertical section

Plan

Purlin top-edge bevel

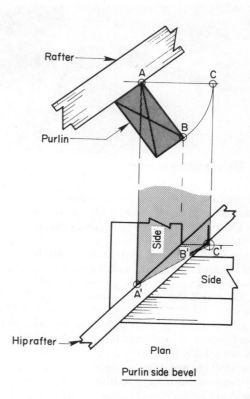

Plan

Purlin side bevel

Fig. 9.6 Purlin geometry

allow the birdsmouth to fit square. An allowance equivalent to the amount removed must be made to the birdsmouth (x), otherwise the rafter will be too long. Hip-rafter overhang is reduced to suit the depth of the common rafters (y).

An angle tie has been fixed across the corner to help prevent outward thrust and to provide extra bearing to the rafter seat cut. Angle ties can be positioned after the rafter is in place.

9.2 Double roof or purlin roof

Purlins are introduced into the roof as a means of mid-rafter support, thereby reducing rafter span by half. Purlins may be fixed vertically or at right angles to the slope of the roof, depending on the type of bearing, for example built-in (preservative-treated), gable walls, or supported by internal (load-bearing) walls either directly or indirectly via corbels, steel hangers, or struts. Roof trusses also offer strutted support.

Where an inclined purlin meets a hip rafter, the intersecting face angles will produce top-edge and side bevels as shown in Fig. 9.6. To find the top-edge bevel, use distance AB to scribe an arc to

meet the horizontal line from A to produce C. Drop a perpendicular line from A on to the plan of the hip to obtain A'. Repeat with B to obtain B'. Draw a horizontal line from B' to meet a perpendicular line drawn from C to produce C'. The angle between AA' and A'C' is the 'purlin top-edge bevel'. Side-edge bevel is found in a similar manner, as shown on the right-hand side of Fig. 9.6.

Note To produce the above bevels it would normally only be necessary to use single lines to represent

a) the pitch of the roof,
b) a top-edge distance of the pitch line,
c) a line at right angles to the pitch line of a length equal to the purlin side,
d) the plan of the hip (45°).

9.3 Steel roofing square

The steel square shown in Fig. 9.7 consists of two arms set at 90° to one another. The 'body' or 'blade' (longest arm) is 600 mm long, the 'tongue' 400 mm. The 'heel' of the square is the corner of the outer edges. Both outer edges are graduated at

Fig. 9.7 'Stanley' steel roofing square

2 mm intervals, reading off in millimetres. Rafter tables are printed on one face, with cuts on the other. (The type of tabulation varies between makes of square.)

Steel squares can be used to obtain both the angle of cut and the length of all the roof members. Different makes vary in their method of application – manufacturer's instructions are supplied with each square.

The geometrical means by which roof bevels and member lengths are obtained may require that roof sizes be scaled down. Scaling down is necessary with the steel square, but only to the extent of determining the number of millimetres the roof rises for every metre run of common rafter (for an equally pitched roof, run = ½ × span).

Figure 9.8 shows that, providing the ratio of mm rise to m run is kept constant, the angles set against the hypotenuse will always be the same. It also follows that the distances along the body (rafter run), tongue (roof rise), and hypotenuse (rafter length) must be proportional to those in the full-sized roof.

Rafter lengths

Reference is made to the tables printed on the body of the square, which are calibrated according to roof rise (mm) per run (m) of common rafter. To obtain this figure, the following calculation will be necessary:

$$\text{millimetre rise per metre run} = \frac{\text{rise (mm)}}{\text{run (m)}}$$

For example, if a double-pitched roof has a rise of 1.5 m and a span of 6 m, its mm rise per m run is

$$\text{mm rise per m run} = \frac{\text{rise (mm)}}{\frac{1}{2} \times \text{span (m)}}$$

$$= \frac{1500 \text{ mm}}{3 \text{ m}} = 500 \text{ mm/m}$$

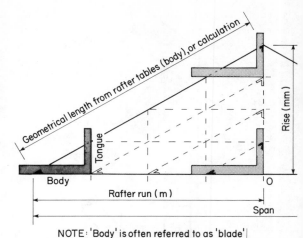

NOTE: 'Body' is often referred to as 'blade'

Fig. 9.8   Proportional mm rise to m run of rafter

In other words, for every metre run the roof rises 500 mm.

Once this figure is found, it can be applied to all the roof members listed on the body to obtain their geometrical lengths.

Alternative methods of finding rafter lengths include

a) calculation, using Pythagoras' theorem or trigonometry;

b) measuring across the hypotenuse of the square and scaling up;

c) physically stepping the square along the rafter ten times (assuming that both the rise and the run had initially been divided by ten to proportionally reduce the size).

Method (a), although accurate, demands mathematical knowledge. Methods (b) and (c) may prove inaccurate – particularly (c), which is prone to error accumulation.

Roof bevels

Figure 9.9 shows theoretically how the diagonal members of the roof relate to the angles shown on the square. However, in practical terms the square has been designed to cope with all the necessary roof bevels simply by using the roof's mm rise per m run marking on the tongue and off-setting this against a 'set mark' (marked by an arrow) on the body to ascertain both the plumb and seat cuts as shown in Fig. 9.10. A purpose-made movable fence (Fig. 9.11) can make setting-up and angle transfers much easier.

With all its tables and figures, the appearance of the steel square can be a little off-putting but, once

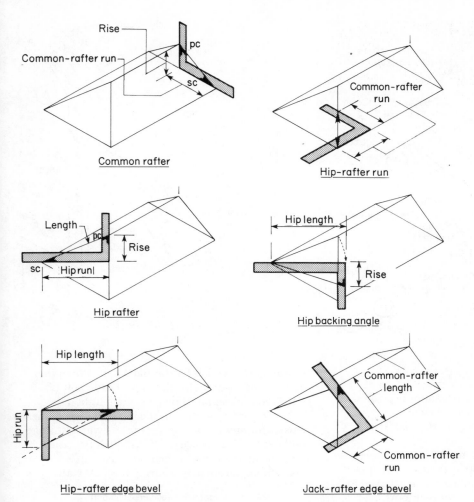

Fig. 9.9 Roof bevels being related to the steel roofing square

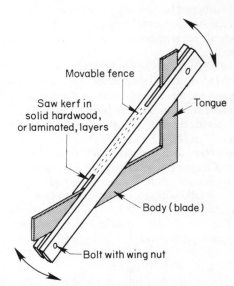

the maker's instruction booklet has been thoroughly read, and after a little practice, it should become much easier to use than first imagined – perhaps then it may have a use other than that of a large try-square.

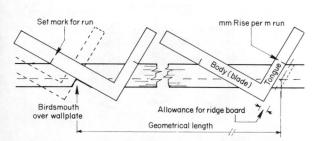

Fig. 9.10 Applying a steel roofing square to produce a pattern rafter

Fig. 9.11 Steel roofing square with movable fence

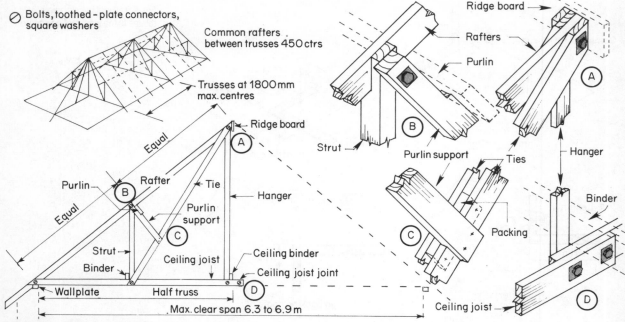

Fig. 9.12 TRADA domestic roof truss

9.4 Trussed roofs

Where suitable purlin supports cannot be achieved from load-bearing walls, or where a clear uninterrupted roof span is required, a roof truss may provide the answer.

Trusses are positioned along the length of the roof at intervals of up to 1.800 m centres to support the purlins which the infilling common rafters – at 450 mm centres – will use as a mid-support. Hipped ends can be formed by using a half truss in place of the crown rafter. Hip rafters are constructed in a similar manner as previously described.

Figure 9.12 shows the construction and assembly of a TRADA domestic-type roof truss. Truss joints in the main require members to be lapped and bolted together with 50 mm double-sided toothed-plate connectors with square washers on the outer sides (Fig. 9.13). The number of connectors between each joint will depend on the number of members being joined; for example, the single lap between rafter and ceiling joist (not shown in detail) will require one connector whereas the joint in Fig. 9.12(a) will require three, that in Fig. 9.12(b) two, that in Fig. 9.12(c) – a nailed joint – none, and that in Fig. 9.12(d) three.

Note Because of the pressure required to drive the teeth of the connectors fully into the wood, a special high-tensile drawing bolt is used to avoid

stripping the teeth off a standard bolt. When the joint is drawn-up tight, the high-tensile bolt is removed and replaced by a permanent bolt.

No provision is made in the initial design of these trusses for any extra loading such as water-storage tanks etc. – if these are to be installed, provision must be made to strengthen the truss and/or to take support from load-bearing internal walls.

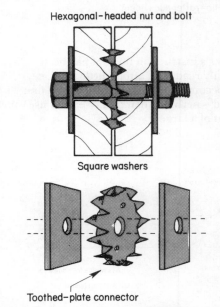

Fig. 9.13 Toothed-plate connection (double-sided connector)

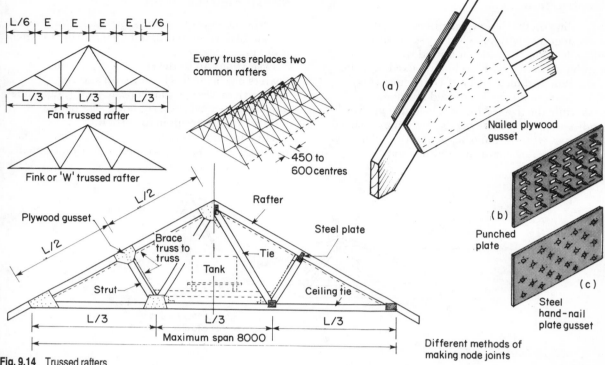

Fig. 9.14 Trussed rafters

Different methods of
making node joints

Fig. 9.15 Unloading trussed rafters

9.5 Trussed-rafter roofs

Trussed rafters have been used in the UK for over 20 years and have now become one of the most acceptable means of providing the roof framework over domestic dwellings.

As shown in Fig. 9.14, each trussed-rafter type (fan and fink) consists of a complete stable framework of struts and ties. The trusses are positioned at 450 mm to 600 mm centres, thereby replacing each pair of adjoining common rafters and their interconnecting ceiling joist. Purlins are dispensed with.

The design of the trussed rafter (responsibility of 'The Trussed Rafter Designer') will depend upon the span, pitch, and loading. One great advantage over traditional methods of construction is that trussed rafters can be pre-made under factory-controlled conditions and be delivered on site (Fig. 9.15) ready to be erected over the wallplates. Site geometry is unnecessary unless hips or valleys form part of the roof plan, in which case traditional means of setting out, as mentioned earlier, will be integrated into the roof.

Construction

Depending on the design, members are either butted together with plywood gussets nailed to both sides (Fig. 9.14(a)), or butted with proprietary galvanised-steel plates fixed over the joints. Punched toothed-plate connectors (Fig. 9.14(b)) are applied with the aid of a machine press while the trussed rafter is held in a jig. Steel hand-nail gusset plates (Fig. 9.14(c)) may be fixed by hand or with the aid of a pneumatic nailer (see Fig. 2.32) or a mechanical hand-nail gun.

Site storage (Fig. 9.16)

Correct stacking of rafters is essential if they are to remain distortion free, and if their moisture content is not to rise above 22%.

Distortion can be the result of supporting bearers not being level – one with another (twisted), or being wrongly positioned (not in-line above another). Trusses banded together too tightly can also result in distortion.

High moisture content is usually due to the stack not being adequately protected against inclement

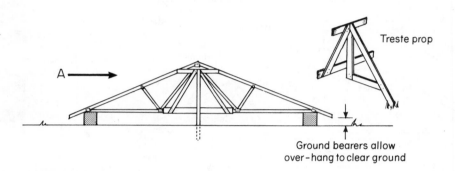

Treste prop

Ground bearers allow over-hang to clear ground

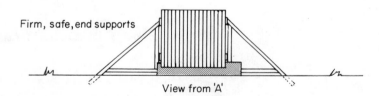

Firm, safe, end supports

View from 'A'

(a) Trussed rafters stacked vertically

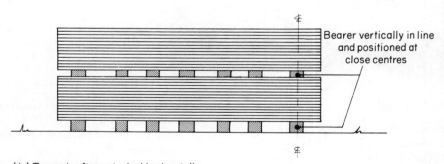

Bearer vertically in line and positioned at close centres

(b) Trussed rafters stacked horizontally

NOTE :– Rafters to be covered to protect from sun, and to shed rainwater-cover to be of the type or positioned in such away as to allow good ventilation to stacks.

Vertical storage preferred, but care must be taken to prevent stack from tipping as trusses are removed.

Fig. 9.16   Site storage of trussed rafters

weather, or by not ventilating the cover and thereby encouraging condensation to develop on the underside of the cover. Ground protection is also necessary to prevent rising damp reaching the trusses or their supports.

Figure 9.16 shows two methods of storage. When trussed rafters have to be stored on site, provision must be made to ensure that they do not become distorted or strained in any way. They should be stored on the ground on raised bearers set parallel with one another. The preferred method is the vertical storing method (Fig. 9.16(a)) where at one end of the supporting framework, the trusses are positioned against a firm and safe support in the form of a trestle prop anchored to the ground, a similar propping arrangement is securely fixed at the other end. Raised bearer supports are provided at each end of the truss (to ensure that the rafter overhang clears the ground) which are positioned at points of support assumed in the design, for example where the wall plates would be.

With horizontal storage (Fig. 9.16(b)) where trusses are stacked flat the bearers must be positioned at every joint to give level support at close centres. Stored trusses should be protected from the weather by a water-proof cover, which must be arranged in such a way to safely allow for the circulation of air – thereby reducing the risk of condensation forming on the underside of the sheeting.

A typical example of a specification for bracing and binders used in the construction of a domestic dual-pitched roof of up to 30°, is described and illustrated in the 'Defect Action Sheet No 83' produced by 'Building Research Establishment'. Set out in Fig. 9.17 is an extract from that document BRE DAS 83.

Stages in roof erection (Fig. 9.18) (spans up to 8 m)

Details should be provided by the 'Building Designer' to suit each particular roof. The following stages in conjunction with Fig. 9.18 should be regarded only as a general guide for domestic roofs.

1 Mark truss-rafter positions on the wallplates, position and fix truss clips (Fig. 9.19) as and when required.
2 Position the first truss rafter to the gable side of where the diagonal (c) would reach the ridge.
3 Once the truss has been plumbed, fix temporary stays (a) to the rafter.
4 Fix the second trussed rafter to the first with temporary batten (b) to the rafters and ceiling ties.
5 Fix the remaining trussed rafters in the same manner.
6 Fix diagonal braces (c) to the underside of the rafters and wall plates.
7 Securely fix all longitudinal members.
8 Fix all remaining longitudinal, diagonal and chevron bracing required on the internal members of the trussed rafters as specified.
9 Fix all lateral restraining anchors, and holding down straps as and where required by the Building Designer and current Building Regulations.

Note

a) All work to be carried out in accordance with the procedures stipulated by the 'Building Designer'.
b) Consult BS 5268 Part 3 : 1985 'Structural use of timber' – 'Code of Practice for trussed rafter roofs'.

Figure 9.20 shows a trussed-rafter roof being erected.

Where water-storage tanks are to be sited in the roof void (Fig. 9.14) and the only means of support are the ceiling ties of the trussed rafters, then two cross-bearers can be laid as close as possible to the one-third (node) point, long enough to span a minimum of three trusses (depending on the size of the tank), with two more bearers over these, then a further two to form the platform. This will ensure that the load is distributed over a wide area. Whenever possible supports for water tanks should be independent of the trussed rafters.

Note With regard to construction, handling, erection, and tank support etc., reference should be made to British Standard code of practice BS 5268 : Part 3 1985 the International Truss Plate Association technical handbook.

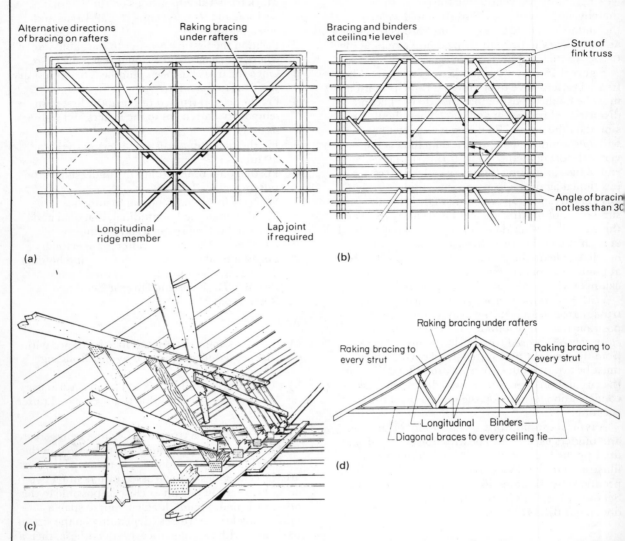

(a)

(b)

(c)

(d)

Alternative directions of bracing on rafters

Raking bracing under rafters

Bracing and binders at ceiling tie level

Strut of fink truss

Longitudinal ridge member

Lap joint if required

Angle of bracing not less than 30°

Raking bracing under rafters

Raking bracing to every strut

Raking bracing to every strut

Longitudinal

Binders

Diagonal braces to every ceiling tie

- Specify, for every roof (or section of roof between cross walls) 100 × 25 mm raking braces, twice nailed to the underside of **rafters** of every truss. The braces should run at approximately 45° from ridge to eaves and be applied to both pitches, (a).

- Specify, when the distance between centres of cross walls is not more than 1.2 × trussed rafter span, at least two 100 × 25 mm diagonal braces, twice nailed to every **ceiling tie** in every roof (or section of roof between cross walls) as shown in (b).
  — where wall spacing exceeds 1.2 × span, specify at least four such diagonal braces in 'W' formation on each side.

- Specify (unless trusses are less than 5 m span), for every roof or section of roof between cross walls, 100 × 25 mm raking bracing twice nailed to every internal strut, (c).

- Specify longitudinal binders, 100 × 25 mm, twice nailed and located as shown in (d);
  — all binders should abut walls at both ends and for this purpose specify each binder to be in two overlapping lengths.
  — where binders cross raking braces the binders should be interrupted and plated, see (c)

- Specify that all lap joints in braces and binders are to be lapped, and nailed over at least two rafters.

- Specify all nailing to be 3.35 × 75 mm galvanised round wire nails.

- Specify that no bracing or binders shall penetrate a separating wall.

Note: Roof pitches up to 30°, spans up to 11 m.

Fig. 9.17 Typical specification details for fixing braces and binders to a trussed rafter roof, by courtesy of B.R.E.

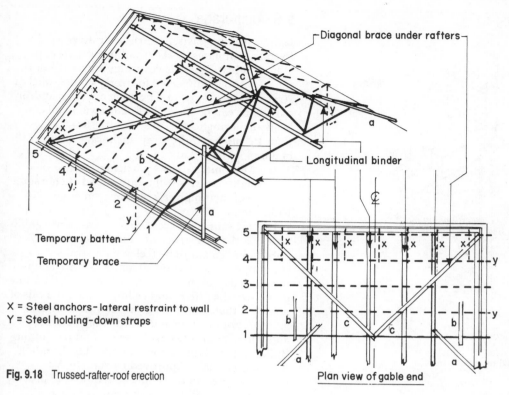

Diagonal brace under rafters

Longitudinal binder

Temporary batten

Temporary brace

X = Steel anchors – lateral restraint to wall
Y = Steel holding–down straps

Plan view of gable end

Fig. 9.18 Trussed-rafter-roof erection

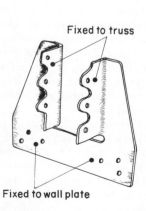

Fixed to truss

Fixed to wall plate

Fig. 9.19 Truss clip

Fig. 9.20 Trussed-rafter roof being erected

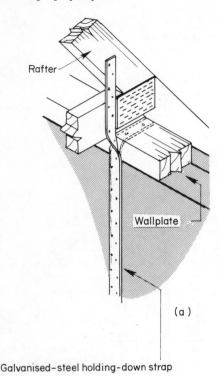

Galvanised–steel holding-down strap

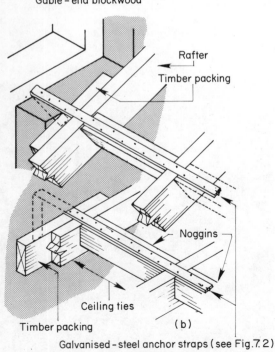

Galvanised–steel anchor straps (see Fig. 7.2)

Fig. 9.21　Lateral support to a gable wall, and roof anchorage

9.6 Anchorage

Rafters, ceiling joists, and trussed rafters will be fixed to wallplates by skew nailing or truss clips. Where lightweight roof coverings are used – or if the roof is to be subjected to strong winds – anchor straps (Fig. 9.21(a)) and framing anchors, and/or truss clips can be used. Wallplates will in all cases need anchoring to the substructure. Roofs of less than 15° pitch require strap anchorage at intervals of not more than 2 m.

Note　Monopitch roofs – particularly those of a lightweight nature – will require good anchorage because of their high risk of displacement due to wind suction.

Lateral support must be given to masonry gable walls via the roof structure. This is achieved by positioning galvanised-steel anchor straps (minimum cross-section 30 mm × 5 mm) against the outer face of the inner masonry leaf and nailing them over the rafters and ceiling ties at intervals of not more than 2 m. Fixings are shown in Fig. 9.21(b). For position and location of anchorage and lateral support consult the current part of the Building Regulations, for example 'The Building Regulations 1985: Approved Document A'.

Note　Large-span roofs may be required to provide lateral wall support along the eaves of the roof.

If gable ends consist of a framework of timber, the roof as a whole must be diagonally braced against the effects of wind.

9.7 Roof openings

Openings may be required for a chimney stack, ventilators, or roof lights (windows).

Figure 9.22 shows how provision can be made to take the roof around a rectangular chimney stack – if the opening is wider than the space between the rafters, two trimmers should be used in place of the noggings to carry trimmed rafters. The surface development of the opening (excluding the perimeter gap) is produced by projecting two lines at right angles to the slope of the roof with a distance apart equal to the sloping length of the opening – the width of the development is equal to *w* on the sectional plan.

Figure 9.23 shows how the true shape (surface development) of an opening produced by a cylinder piercing a sloping roof is obtained and how that portion of the cylinder protruding above the roof would look if it were opened out 'flat'.

To develop the opening, mark points a to g equally spaced around the circumference (plan) of the cylinder and project them up to the pitch line

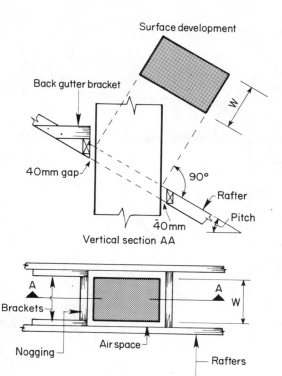

Fig. 9.22 Rectangular chimney stack or duct piercing a pitched roof

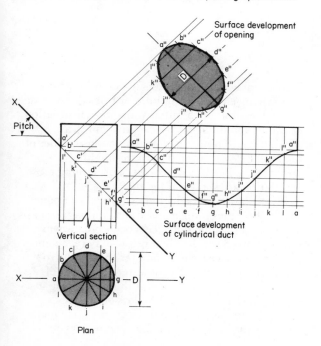

Fig. 9.23 Cylindrical duct piercing a pitched roof

(vertical section XY), to obtain a′, b′, c′, etc. Project lines from points a′ to g′ at right angles to the pitch line and, at a convenient point, draw a line a″g″ at right-angles across them. From centre line ag on the plan, transfer vertical distances bl, ck, dj, ei, fh, to the surface development to produce b″l″, c″k″, . . . and so on. The points are then joined together freehand to produce the elliptical-shaped development.

To develop the cylindrical projection, project a′ to g′ horizontally. Vertical lines which form part of the grid are spaced at intervals equal to the arc lengths ab, bc, cd, . . . and so on. Where these lines meet, they give points a″, b″, c″, . . . and so on. Join these points together freehand to produce the development.

Note   Openings to be formed around flues, whether at roof or ceiling level, must comply with current Building Regulations with regard to proximity of combustible materials.

An opening in a ceiling used as a manhole for access to services etc. may – depending on its size and joist centres – require trimming (use of joist hangers etc.). Its edges should be lined out to accommodate a hinged trap door which, for safety reasons, is made to open into the 'loft' (roof void), the joint between the lining and plasterboard being covered with a 'trim' (architrave). If a loft ladder is to be installed then the trap door usually forms part of the arrangement and is therefore hinged to open into the room or landing below.

9.8 Thermal insulation and ventilation

As stated in Section 8.8, thermal insulation must be provided to satisfy current Building Regulations. In the case of a pitched roof with a ceiling, insulation is usually in the form of a quilt sited in between ceiling joists. A vapour barrier can be positioned on the underside of the insulation – foil- or polythene-backed plasterboard may be suitable, provided the joints are sealed – this barrier can help reduce the amount of water vapour which would normally pass through the ceiling into the roof void.

Figure 9.24 shows how the insulation within the eaves of this traditionally constructed roof is taken on to the inner leaf of the wall at eaves level – a board prevents the insulation from blocking air flow conducted via a vented soffit. Ventilation is vitally important in the roof void if condensation is to be avoided. Means of providing ventilation include a continuous insect screened eaves opening along two opposite sides of the roof (Fig. 9.25 shows a proprietary type of eaves ventilator used

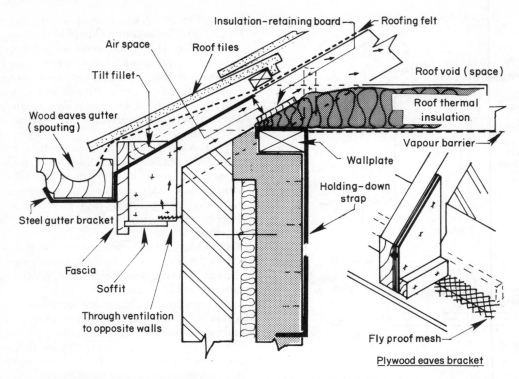

NOTE: Wood eaves gutter (still preferred by many joiners) is still available, but the use of plastics or metal gutter is more common on new developments.

Fig. 9.24   Insulation and eaves detail (traditional roof construction)

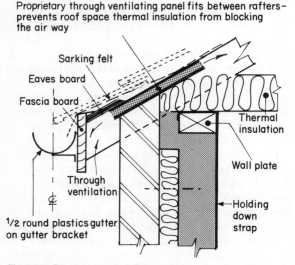

Fig. 9.25   Proprietary roof ventilator used in conjunction with open eaves

with a roof with 'Open Eaves'). Gable vents may accompany eaves ventilation (air bricks), and ridge vents. For roofs of 15° pitch and above, ventilation openings should be of a size equivalent to a continuous 10 mm gap along two opposite sides of the roof; below 15° pitch the gap should be increased to 25 mm. Examples of promoting cross ventilation are shown in Figure 9.26.

If ventilation is inadequate and/or the vapour barrier becomes inoperative due to damage or gaps left around the edges of access traps etc., warm moist air from the dwelling below may reach and make contact with the cold outer shell of the roof, condensing into droplets of water – possibly of sufficient volume to cause continuous wetting to timber, which could result in loss of timber strength, fungal decay, and the corrosion of metal components such as truss plate connectors etc.

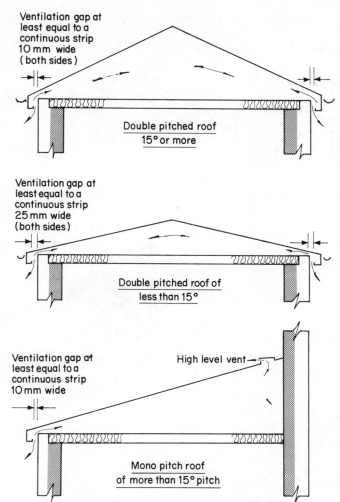

Ventilation gap at least equal to a continuous strip 10 mm wide (both sides)

Double pitched roof 15° or more

Ventilation gap at least equal to a continuous strip 25 mm wide (both sides)

Double pitched roof of less than 15°

Ventilation gap at least equal to a continuous strip 10 mm wide

High level vent

Mono pitch roof of more than 15° pitch

Fig. 9.26 Method of promoting cross ventilation to a roof space

9.9 Eaves gutters and down pipes (fall pipes)

Eaves gutters

As mentioned and shown in Figure 9.24 wood eaves gutters are still quite popular in some areas. Butt and mitre joints (Fig. 9.27) usually incorporate a half-lap, this not only strengthens the joint but ensures alignment between adjoining lengths. Joints are usually sealed with a bituminous mastic compound, then covered with a strip of lead which should be recessed flush with the channel surface. After the lead has been secured to the gutter with closely spaced copper tacks the heads are covered with a thin film of the mastic – lead funnels (Fig. 9.27) which act as water outlets are pushed through holes cut through the channel of

the gutter – an expansive bit (Book 1 Fig. 2.57) is used to bore the hole through the gutter. Holes should be bored from the outer face of the gutter. Using a scribing gouge and mallet funnel, flanges are recessed and then fixed in the same manner as above. The whole of the inner surface of the gutter is treated with a water repellent (traditionally bituminous paint or tar) paint.

Note Lead joints may be the job of the plumber.

Plastics are now the most common gutter material – available in a variety of section sizes and lengths. Plastics gutter must be well supported by using the appropriate fascia or rafter bracket fixed at centres not less than those specified by the manufacturer. Joints must be sealed in accordance to the manufacturer's instructions. Gaps between joints are left to allow for thermal movement, for

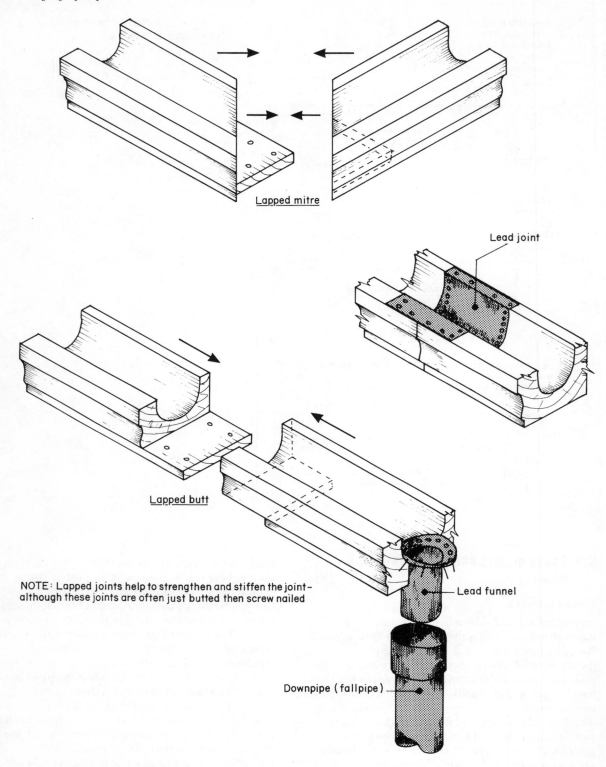

Fig. 9.27   Traditional wooden gutter joints and connections

example, a gutter length of up to 2 m which is to be assembled during the winter months should have a joint gap of 7 mm, if the same job was carried out during the summer months than a joint gap of 5 mm should be left. Gutter lengths between 2–4 m will require gaps of 10 mm in winter and 7 mm in summer. Failure to leave expansion gaps could result in the gutter becoming distorted or buckled. Plastics components such as guttering are usually fixed by the plumber. Metal gutters for domestic property are generally supplied and fixed by firms who specialise in that type of work.

All eaves gutters should have sufficient slope to ensure that all the water likely to be discharged into them from the roof should run away without leaving any puddles. The direction of fall will depend upon the location of the nearest down pipe (fall pipe). Domestic eaves gutters should have a minimum fall of 1 in 350 or about 10 mm in 3 m.

Alignment of the eaves tiles/slates with the gutter (channel) centre line as shown in Fig. 9.25 is important as this not only ensures that roof water enters the gutter correctly, it makes adequate provision for the eaves sarkin felt to enter the gutter back, and provides for reasonable access to gutter channels for cleaning and maintenance purposes.

Down pipes (fall pipes)

Plastic is now the most common material for down pipes. Adequate wall fixings are important, distances between fixing should be specified by the manufacturers and these distances should never be exceeded. As with gutters, provision for thermal movement must be made – this can be achieved by pushing home the joint, then slightly withdrawing the spigot from the socket by the amounts previously stated for gutters.

10

In-situ formwork

In Book 1, formwork (shuttering) was divided into two areas of work: (i) in-situ formwork and (ii) pre-cast work involving the use of mould boxes to produce transportable concrete units.

This chapter deals solely with work which may be found in situations similar to those shown in Fig. 10.1.

To conform to BS 6100 (Sec. 6.5): 1987, the term 'formwork' will be used throughout the text in place of the trade term 'shuttering'.

10.1 Formwork design

Formwork should be designed and suitably constructed to be capable of

a) withstanding dead and live loads,
b) being easily assembled and dismantled,
c) accurately producing or reproducing concrete items of correct size and shape,
d) producing the desired finish on the concrete face,
e) reusability – for reasons of economy.

Loads on formwork

Dead loads are static loads such as the weight of the formwork and wet concrete. The dead load on the formwork due to the concrete and its steel reinforcement will become less as the concrete hardens and becomes self-supporting.

Live loads are those which are imposed during the erection of the formwork and the placing of concrete. They include the movement of the workforce, their materials, and equipment.

Unlike formwork used for suspended floor or roof slabs, which in the main is subjected to vertical loads, formwork for columns, walls, and deep beams will have to resist varying degrees of lateral pressure as a result of the build-up of concrete in its fluid state (fluidity is increased during its compaction by vibration) within the confines of the formwork.

Fluid pressure exerted on the formwork can be compared to the pressure built-up under a head of water, known as 'hydrostatic' pressure. Figure 10.2 shows how the head of water affects the pressure exerted on the cylinder walls at various levels – the same principle will apply to columns, walls, and deep beams.

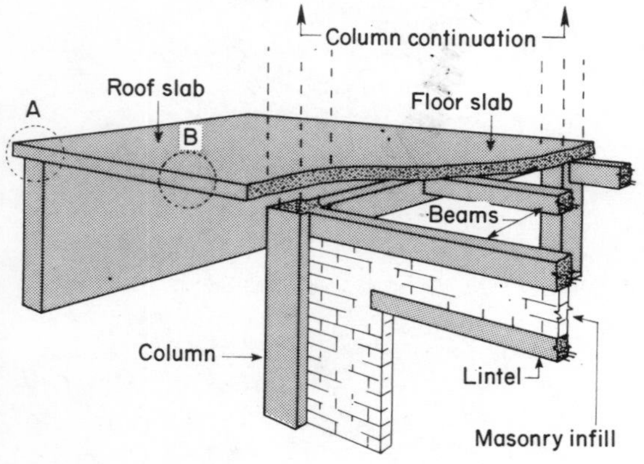

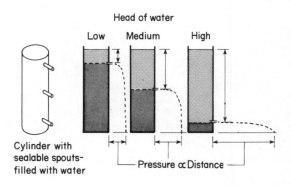

Fig. 10.1 Key to in-situ-concrete structural elements

Fig. 10.2 The effects of hydrostatic pressure

Form (shutter) support and formwork hardware

'Forms' consist of those parts of the formwork that come into contact with the concrete (sheeting) together with their stiffening members, and they are often made-up into panels. They will not be strong enough to withstand the loads and pressures mentioned without added support.

Forms for suspended floor and roof slabs require an arrangement of evenly distributed 'ledgers' (horizontal members supported by props or steel hangers) and either props of timber with hardwood folding wedges for adjustment or adjustable steel props (Fig. 10.12), to give support to the joists and decking (form) against vertical loads.

Figure 10.3 shows how a column-box, wall, or deep-beam form might react to the lateral pressures of the wet concrete if only top and bottom restraints were used. In all cases it will be seen that greater restraint will be required towards the base. In the case of column forms, various types of 'yoke' (a clamping arrangement fixed around the forms to restrain lateral pressures) and raking props are used. Wall forms use 'form ties' (devices for restraining both side forms), 'walings' (horizontal members), and/or 'soldiers' (vertical members) as well as raking props. Beam forms may take their

support from 'beam clamps' (vice mechanisms supported by adjustable steel props), 'head trees' (strutted bearers on top of props) and props, or suspended steel hangers.

Sheeting (sheathing) and form lining

As mentioned above, sheeting makes contact with the concrete. Form linings also contact the concrete but refer to those materials applied or fixed to the inside face of the form to give the face of the concrete a special finish.

Sheeting materials include boards of timber, but in the main these have been superseded by sheets of exterior (formwork-grade) plywood. Moisture-resistant chipboard and oil-tempered hardboard are other alternatives which, like plywood, can produce an almost unblemished joint-free finish on the concrete. Steel is used either as an all-steel panel system or as a form framework for plywood sheeting. Forms made-up of glass-fibre-reinforced plastics (GRP) which have been pre-fabricated under factory conditions can be used where complex shapes have to be cast.

The above materials may also be used as a form lining to produce finishes ranging from the very smooth, through the textured, to those with special patterned features. Best results are obtained by coating the lining with plastics solutions or film overlay – their impervious nature gives good release qualities (reduces adhesion). However, some form linings will remain permanently in place after the formwork is struck, to provide the surface of the concrete with either acoustic or thermal-insulation properties – fibreboard and foamed plastics are two materials which may be considered for this purpose.

Preparing forms for concrete

Screw and nail holes which may affect the finish should be filled with a waterproof stopper. Sheeting and lining cut edges should be sealed, and their surfaces should be clear of dirt, sawdust, chippings, or any foreign matter which could effect the 'keying' (adhesion to kickers and reinforcement) and setting qualities of the concrete. Deep formwork may require a purpose-made clean-out hole (Fig. 10.4(c)).

To prevent concrete sticking to the surface of the forms, a parting or release agent is used. Release agents may be in the form of an oil (mould oil), an emulsion, or a synthetic resin or plastics compound – the latter two may also require oil treatment. Treatment of this nature also increases the life of

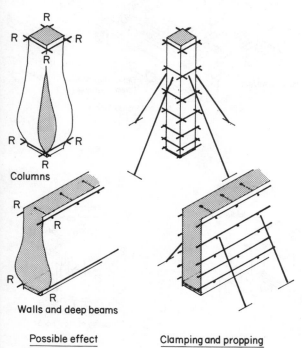

Columns

Walls and deep beams

| Possible effect | Clamping and propping |

R = Restrained

Fig. 10.3 Resisting the effect of hydrostatic pressure

the forms (reusability) and helps reduce the number of surface blemishes and 'blow-holes' (holes left by pockets of air) appearing on the finished concrete.

Some treatments are not suitable for metal forms, as they tend to encourage rusting – maker's recommendations should always be observed with regard to use and application. Application methods may include spraying, but mould oils are more usually applied by swab or brush.

Note   Care should be taken not to get mould oils, or any release agents, on to steel reinforcement or areas where adhesion is important.

10.2 Column formwork

Figure 10.4 shows various stages and arrangements for constructing formwork suitable for a square or rectangular column.

Formwork must be securely anchored at its base, to ensure that it does not move from its appointed place during its construction or while the concrete is being poured. This can be achieved by one of two ways: by using formwork as shown in Fig. 10.4(a) to cast a kicker (a short concrete upstand above floor level, around which the formwork above is anchored) or by fixing an anchor framework to the floor (Fig. 10.4(b)) into which the whole column box sits and is securely fixed to it.

Figure 10.4(c) shows the bottom section of a column box. Its boarded forms are held in place by traditional timber yokes and double-ended steel bolts and hardwood wedges. A clean-out hole with a detachable door is also shown.

Yokes made up of adjustable steel column clamps are shown in Fig. 10.4(d). They have been assembled in a staggered manner so that, when the wedges are tightened, they do not twist the formwork. Nails are used at all clamp location points, as a means of holding the clamps in position

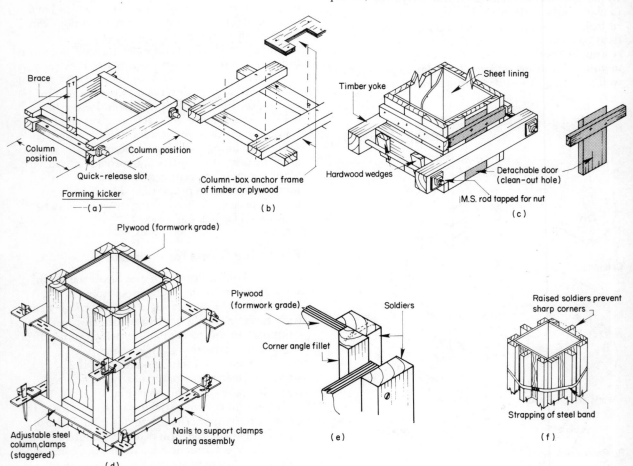

Fig. 10.4 Column-box construction and anchorage

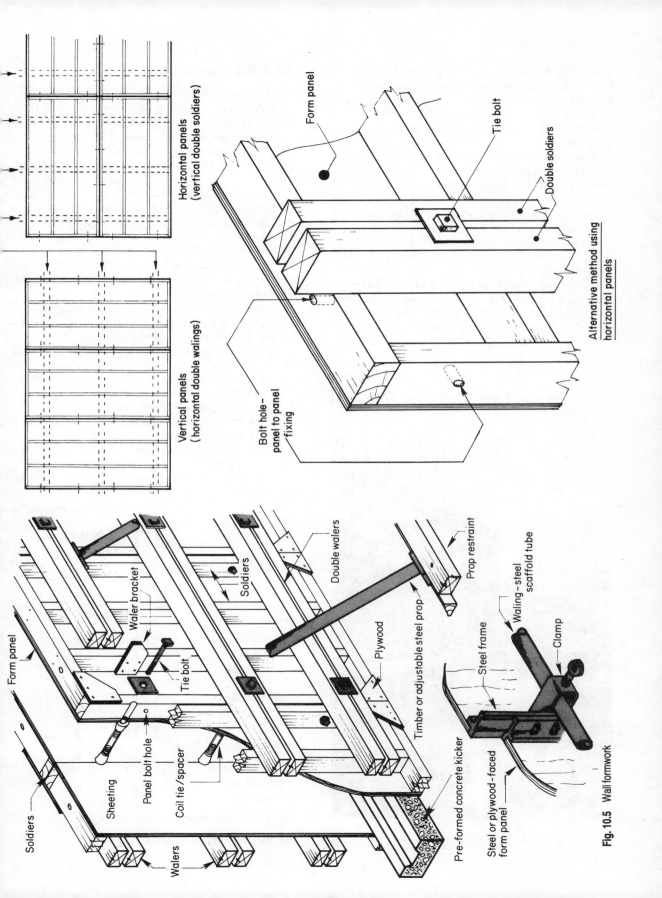

Horizontal panels
(vertical double soldiers)

Vertical panels
(horizontal double walings)

Form panel

Tie bolt

Double soldiers

Alternative method using
horizontal panels

Bolt hole –
panel to panel
fixing

Form panel

Soldiers

Waler bracket

Tie bolt

Soldiers

Double walers

Sheeting

Panel bolt hole

Coil tie/spacer

Plywood

Timber or adjustable steel prop

Prop restraint

Pre-formed concrete kicker

Steel or plywood-faced
form panel

Steel frame

Waling – steel
scaffold tube

Clamp

Soldiers

Walers

Fig. 10.5 Wall formwork

while they are being tightened. Figure 10.4(d) shows plywood sheeting being used – corner fillets prevent sharp corners, which otherwise tend to become damaged on striking and while the concrete is green. Joint details are shown in Fig. 10.4(e).

A steel strap has been used as a yoke in Fig. 10.4(f). Yokes of this type are applied in a similar manner to straps used to bind packing cases. Additional soldiers have been introduced midway along each side, to avoid sharp corners and provide extra bearing.

Forms can be temporarily nailed together in pairs to form an 'L' shape – two of these pairs then form the column box. After the yokes have been fixed, the formwork is plumbed and held that way with raking props (Fig. 10.3). Prop heads should rest against form bearers – not the yokes – and feet should be restrained from sliding (Fig. 10.5).

Fig. 10.6 Erection of a wall form

10.3 Wall formwork

Wall forms are usually made up of prefabricated units or panels. A typical formwork assembly is shown in Figs 10.5 and 10.6 and includes soldiers, walings, props, form ties, bolts, and clamps.

Some proprietary form ties and spacers are shown in Fig. 10.7. The 'coil' tie (Figs 10.7(a) and 10.8) is made from two coils of steel bar, held apart by two short lengths of bar welded to them. The coils receive the thread of the bolts. Coil ties remain embedded in the concrete. The plastics cone ends and tie also serve as a distance piece, to keep both forms the correct distance apart. After the formwork is struck, the cones are removed and their holes are either filled with fine concrete or left as an anchor point (Fig. 10.8).

A 'tapered' tie (Fig. 10.7(b)) is a through tie – i.e. the whole tie is withdrawn after the formwork is struck. Because of its taper, it can easily be withdrawn by applying pressure with a spanner to the wide end of the tie in the direction shown. Once the nuts/washers are fixed to the forms, a tapered tie also becomes a distance piece.

The centre portion of the 'crimped' tie (Fig. 10.7(c)) remains in the concrete. The end pieces are withdrawn as shown and the holes are filled.

Figure 10.7(d) shows basically a long double-ended bolt which passes through a tube and two cones (distance pieces). The tube remains embedded.

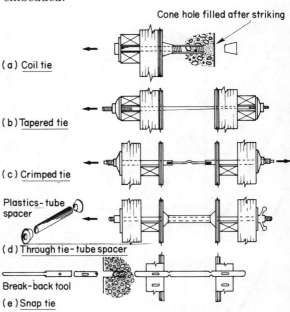

Cone hole filled after striking

(a) Coil tie

(b) Tapered tie

(c) Crimped tie

Plastics-tube spacer

(d) Through tie-tube spacer

Break-back tool

(e) Snap tie

NOTE: Ties withdrawn in the direction of the arrow

Fig. 10.7 Form ties and spacers

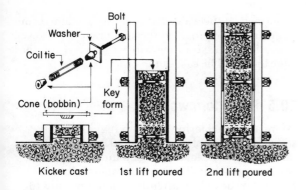

NOTE : Props and ground anchors not shown

Fig. 10.8 Climbing formwork

'Snap' ties (Fig. 10.7(e)) are used with steel forms. As the name implies, the ends of the ties are snapped off, using a special tool, leaving the break within the concrete. The holes are filled.

Where, because of their height, walls are to be cast with two or more lifts of formwork, an arrangement similar to Fig. 10.8 could be used. A kicker is cast and, via the retained coil ties, is used as anchorage for the first lift. Subsequent lifts similarly use previous lifts as anchorage. Progression in this manner is termed 'climbing formwork'.

Note Clean-out holes may be required at the base of column and wall formwork.

10.4 Suspended floor and roof formwork

Decking units (forms) are usually prefabricated into standard sizes. These framed panels, made up of joists and decking, are positioned on to a pre-arranged assembly of ledgers, props, and bracing as shown in Fig. 10.9(a) or are positioned by crane as a table form (Fig. 10.10). Provision

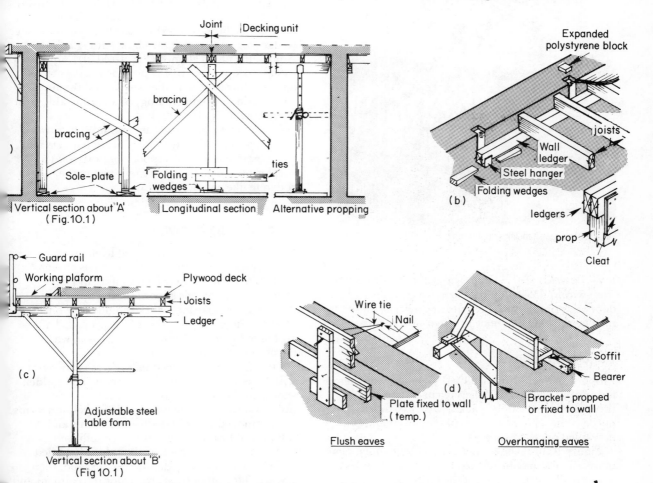

Fig. 10.9 Formwork for a suspended floor or roof slab (Note: section sizes of members will depend on loading span etc.)

Fig. 10.10   Crane handling a table form

must be made for easy removal of all the formwork from within the structure – sizes of openings and doorways should be taken into account, as these may be the only exit once the floor or roof is cast.

Edge treatment

Props or hangers (Fig. 10.9(b)) may be used where forms abut walls. The concrete cut-out on which the hangers rest can be packed with expanded polystyrene which can be easily removed and the hole filled once the slab is cast.

Figure 10.9(c) shows how slab edges to openings or overhangs may be formed (this detail includes a working platform as part of the formwork).

Figure 10.9(d) shows how eaves and verges can be formed. If the walls had been cast, retained coil ties could have provided the anchorage for brackets or soldiers. The straining wire used to restrain the edge form will be cast in the concrete.

10.5 Beam formwork

Beam and lintel formwork both require side and soffit support, as shown in Fig. 10.11. Because it takes most of the load, the soffit and its supports will be struck last. To avoid members becoming trapped or boxed-in, attention should be paid to the sequence of striking, particularly when beam and slab formwork are as one. An example of how a decking may become trapped within a beam side is shown in Fig. 10.11(a).

Figure 10.11(b) shows a soffit form sat on a head tree made up of a timber bearer sat across two ledgers, propped by an adjustable twin steel prop. The stirrup heads are wedged cornerwise to ensure that the bearing is centralised over each leg. Alternative head trees which use hardwood folding wedges for adjustment are shown in Fig. 10.11(c). Adjustable beam clamps (Fig. 10.11(d)) supported by steel props (Fig. 10.12) provide support to soffit and side forms which otherwise would require either a timber strut (raking) or an adjustable steel beam strut (Fig. 10.11(a)).

If the concrete is to be cast around structural steelwork (Fig. 10.11(e)), steel beam hangers may be a possible means of support for both soffit and side forms. The hangers are retained in the concrete like coil ties.

10.6 Striking formwork

Formwork must not be removed from any structural concrete until the concrete is strong enough to carry both its own weight and any weight which may be superimposed upon it. Approval to strike formwork should come only from the engineer in charge of the project. Table 10.1 should only be used as a guide as to when forms may be available for reuse.

Striking must be a gradual process carried out with care to avoid giving the concrete any sudden shock. In the stages of formwork construction, provision should have been made for column forms to be removed without disturbing beams and soffits, and beam side forms before soffits etc.

As each form is struck, it should be de-nailed and cleaned, with any damage made good, then stacked flat (out of twist) ready for retreatment and reuse.

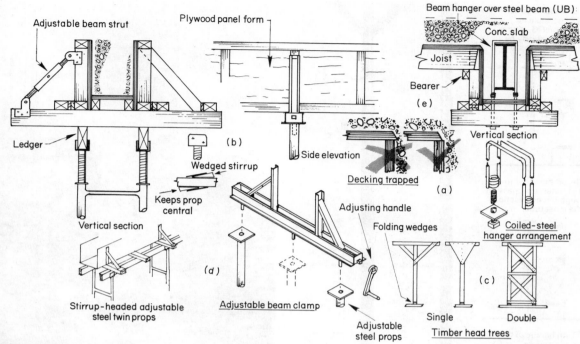

Fig. 10.11 Beam and lintel formwork

Table 10.1 Guide to minimum period before striking formwork for Ordinary Portland Cement concrete (BS 8110 : part 1 : 1985)

Type of formwork	Surface temperature of concrete	
	16°C and above	7°C
Vertical formwork to columns, walls, and large beams	12 hours	18 hours
Soffit formwork to slabs	4 days	6 days
Soffit formwork to beams and props to slabs	10 days	15 days
Props to beams	14 days	21 days

Figure 10.13 shows a column which has just been struck, and a further two with props removed waiting to be struck. Figure 10.14 shows pieces of formwork at different stages of construction – notice that hard hats are being worn by the operatives.

(A) Lift inner tube as near as possible to height required. (Outer tube is kept steady by placing foot on base plate.)

(B) Insert pin through slot in outer tube, passing through the nearest hole in the inner tube.

(C) Turn handle of nut for final adjustment.

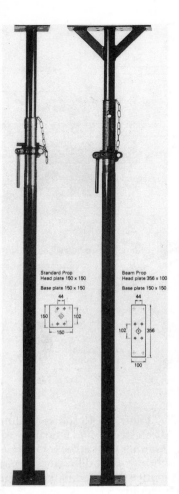

Standard Prop
Head plate 150 x 150

Base plate 150 x 150

Beam Prop
Head plate 356 x 100

Base plate 150 x 150

Fig. 10.12　'Acrow' adjustable steel props

Fig. 10.13　Concrete columns

Fig. 10.14　Formwork at different stages of construction

11

Scaffolding

A scaffold can be regarded as a temporary structure which provides a safe working platform for operatives and their materials. It may consist of or be made up of ladders, trestles, or tubular metal (steel or aluminium) with a platform of wood or metal.

The more common forms of scaffold broadly fall into the following groups:

a) ladder scaffold,
b) trestle scaffold,
c) putlog tied scaffold (often called a bricklayer's scaffold),
d) independent tied scaffold,
e) system scaffold,
f) mobile scaffold.

All scaffolding and the means of getting on and off it must be constructed and maintained in accordance with the provisions laid down in the Construction (Working Places) Regulations 1966; together with the Health and Safety at Work Act 1974, require that those persons involved in erecting scaffolding are 'competent' to do so. A competent person could be regarded as one who possesses sufficient practical and theoretical knowledge, together with enough actual work experience, to be capable of carrying out the job in question safely and efficiently – which would include being familiar with all relevant statutory requirements and the contents of the following British Standards; and guidance notes from the Health and Safety Executive;

BS 5973:1981	Code of practice for access and working scaffolds and special scaffold structures in steel
BS 1139	Metal scaffolding
part 1:1982	Specification for tubes for use in scaffolding
part 2:1982	Specification for couplers and fittings for use in tubular scaffolding
part 3:1983	Specification for prefabricated access and working towers
part 4:1982	Specification for prefabricated steel splitheads and trestles
BS 1129:1982	Timber ladders, steps, trestles, and lightweight stagings
BS 2037:1984	Aluminium ladders, steps, and trestles for the building and civil engineering industries
BS 2482:1981	Specification for timber scaffold boards
HSE	Guidance note GS 15 General access scaffolding
HSE	Guidance note GS 31 Safe use of ladders, step-ladders and trestles
HSE	Guidance note GS 42 Tower scaffolds

11.1 Ladders

See: Health & Safety Guidance Notes GS 31. 'Safe use of ladders, step ladders and trestles'.

A ladder is either used independently to reach a job of work or, as part of a scaffold, is used to allow operatives to reach or leave the working platform. It may in some cases offer support to a platform (ladder scaffold).

There are three classes of ladder as stated in BS1129 and 203.

Class 1 *Industrial (heavy duty)*
High frequency and onerous conditions of use. Carriage and storage suitable for industrial use.
Class 2 *Light trades*
Medium duty, low frequency and good conditions of use, carriage and storage. Suitable for light trade purposes.
Class 3 *Domestic*
Light duty. Suitable for domestic and household use only.

The class and duty must be clearly marked on the equipment. Many accidents happen due to the wrong class of ladder being used. Figure 11.1 shows four different types of timber ladders (with the exception of the 'pole' ladder, these may also be constructed of aluminium alloy). Minimum

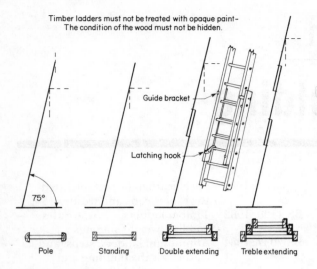

Timber ladders must not be treated with opaque paint—
The condition of the wood must not be hidden.

Guide bracket

Latching hook

75°

Pole   Standing   Double extending   Treble extending

Fig. 11.1   Ladder types

recommended overlaps for timber extending ladders are as shown in Fig. 11.2. Suitable timber species for use in ladder construction are listed in Table 11.1.

Methods of raising, lowering, and carrying ladders are shown in Fig. 11.3. Unless only short ladders are involved, all these operations require two operatives – one to act as anchor, the other to raise or lower the ladder. By employing two

persons to carry the ladder, the load is halved and manoeuvring around corners and obstacles is made safer and simpler.

Figure 11.4 shows various means by which the top and bottom of a ladder can be prevented from slipping. Where a ladder is used to reach a landing (working platform), it must extend above it by not less than 1.07 m, to provide a handhold when stepping on to and off the landing. The slope of the ladder should be at an angle of about 75° to the horizontal – set one unit out to every four units up.

11.2 Ladder scaffold

(Fig. 11.5) See: Health & Safety Guidance Notes GS 31. 'Safe use of ladders, step ladders and trestles'.

Ladders used for ladder scaffolds must be strong enough to support the operative and working platform.

Guard rails and toe boards as such are not necessary if a handhold is provided throughout the whole length of the platform. The width of the platform should, unless impracticable, be at least 432 mm. Separate ladder access should be used to mount the working platform of ladder scaffold. *Note*   This type of scaffold must only be used in connection with work of a light nature and of short duration where such work can be carried out safely.

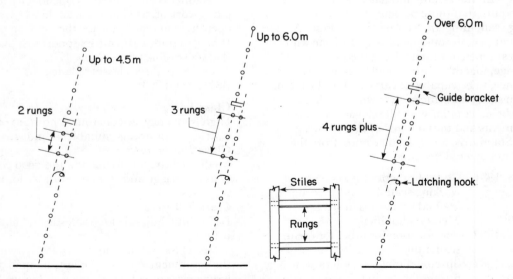

Up to 4.5 m

2 rungs

Up to 6.0 m

3 rungs

Stiles

Rungs

Over 6.0 m

Guide bracket

4 rungs plus

Latching hook

NOTE: Secure as necessary to minimise undue swaying and sagging

Fig. 11.2   Minimum rung-over-rung overlap for extending ladders (guide only)

Table 11.1 Suitable timber species (indicated by a '√') for ladders, steps, trestles, and lightweight staging

Item	Component	Hardwoods										Softwoods					
		European oak*	American white oak*	European ash*	American white ash*	Hornbeam	Yellow birch	Hickory*	Robinia	Keruing	Ramin	European whitewood	European redwood	Douglas fir	Sitka spruce	Western hemlock	Parana pine
Pole ladder	Stiles											√	√				
	Rungs	√	√	√	√	√	√	√	√	√	√						
Standing and extending ladders	Stiles											√	√	√	√	√	
	Rungs	√	√	√	√	√	√	√	√	√	√						
Step ladder	Stiles											√	√	√	√	√	
	Tread									√	√	√	√	√	√	√	√
Trestles	Stiles											√	√	√	√	√	
	Cross-bearers	√	√	√	√	√	√	√	√	√	√	√	√	√	√	√	
Lightweight staging	Stiles											√	√	√	√	√	
	Cross-bearers	√	√	√	√	√	√	√	√	√	√	√	√	√	√	√	
	Decking											√	√	√	√	√	√

Note The same timber species must be used for all components of any one item.
*Check growth rate

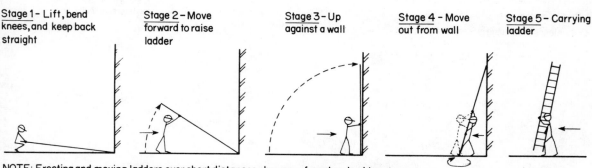

Stage 1 – Lift, bend knees, and keep back straight

Stage 2 – Move forward to raise ladder

Stage 3 – Up against a wall

Stage 4 – Move out from wall

Stage 5 – Carrying ladder

NOTE: Erecting and moving ladders over short distances – beware of overhead cables etc.
Metal ladders must not be used in areas where electric cables are present

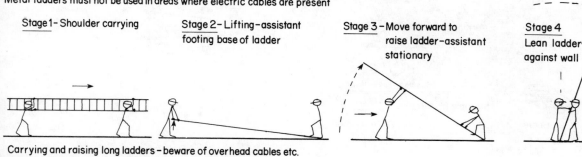

Stage 1 – Shoulder carrying

Stage 2 – Lifting – assistant footing base of ladder

Stage 3 – Move forward to raise ladder – assistant stationary

Stage 4 Lean ladder against wall

Carrying and raising long ladders – beware of overhead cables etc.

NOTE: Each section of extending ladders should be raised or lowered separately – fixed sections (pulley-roped etc.) in their closed position. If the ladder is being carried by one person on the shoulder then the forward end must point towards the ground

Fig. 11.3 Manipulating and manoeuvring ladders

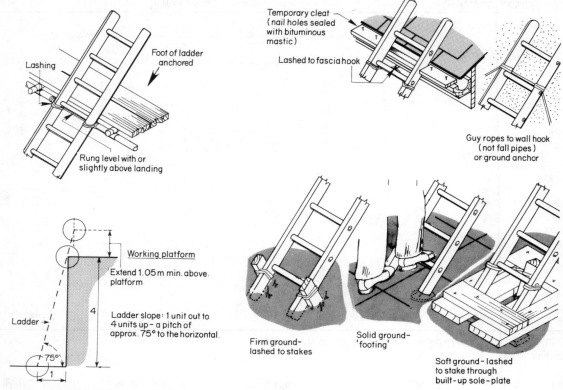

Lashing

Foot of ladder anchored

Rung level with or slightly above landing

Temporary cleat (nail holes sealed with bituminous mastic)

Lashed to fascia hook

Guy ropes to wall hook (not fall pipes) or ground anchor

Working platform

Extend 1.05 m min. above platform

Ladder slope: 1 unit out to 4 units up – a pitch of approx. 75° to the horizontal.

Ladder

75°

4

1

Firm ground – lashed to stakes

Solid ground – 'footing'

Soft ground – lashed to stake through built-up sole-plate

Fig. 11.4   Means of securing ladders

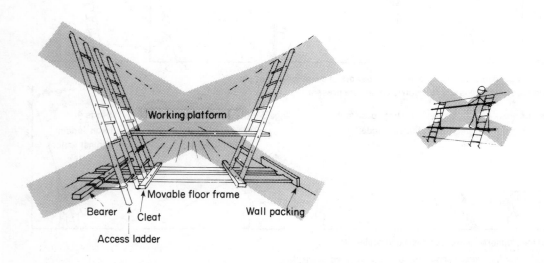

Working platform

Movable floor frame

Wall packing

Bearer   Cleat

Access ladder

Fig. 11.5   Ladder scaffold. Only industrial type ladders to be used capable of carrying the loads imposed

11.3 Trestle scaffold

See: Health & Safety Guidance Notes GS 31 'Safe use of ladders, step ladders, & trestles'.

A trestle scaffold consists of a working platform (staging) supported by two or more trestles. Folding steps (Figs 11.6(a) and (b)) are often suitable as a means of access to trestles for low-rise work. 'A-frame' trestles (Fig. 11.6(c)) which may be used to operate at higher levels are often called 'painter's' or 'shopfitter's' trestles. A scaffold arrangement, together with its limitations, is shown in Fig. 11.7. The need for check blocks, which prevent the hinges from becoming strained when the trestles are being carried or stored, is also illustrated. In the open position ropes may be used to prevent strain on hinges.

Rigid trestles (Fig. 11.8) are available either as fixed-width units or as individual legs (splitheads) with stirrup heads designed to receive a single (as shown) or double beam which allows two platforms to butt together. By using multiples of these trestles, large flat platformed areas can be constructed. Toe boards and guard rails are required at all heights exceeding 2 m.

No trestle scaffold should be used where a person would be liable to fall more than 4.5 m from its working platform. Folding supports should never accommodate more than one working platform. The platform (not less than 430 mm wide) may be made up of either scaffold boards or a proprietary type of staging similar to that shown in Fig. 11.9. A separate ladder or pair of steps should be used for access to the platform if its height exceeds 2 m.

Suitable timber species for the construction of folding step-ladders, trestles, and lightweight staging, are listed in Table 11.1.

(a)

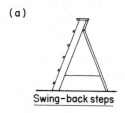

Swing-back steps

(b)

Handhold →

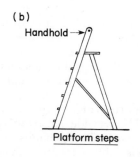

Platform steps

(c)

Staging

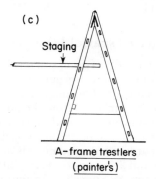

A-frame trestlers (painter's)

Note: Timber equipment (steps, trestles etc) must not be treated with an opaque paint- condition of the wood must not be hidden

Fig. 11.6 Folding steps and trestles

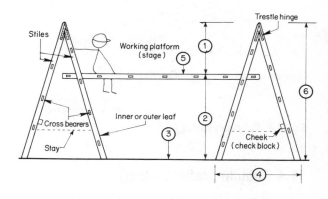

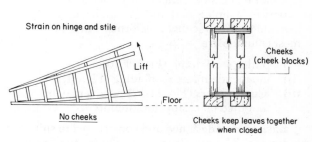

Fig. 11.7 Folding-trestle scaffold 1 – upper third of trestle above working platform 2 – must not be used where a person could fall more than 4.5m; materials must not be placed on the scaffold 3 – firm level ground 4 – trestles fully opened 5 – proprietary staging – recommended 6 – trestles higher than 3.6m should be tied to the building

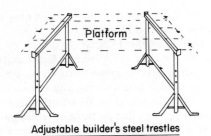

Adjustable builder's steel trestles

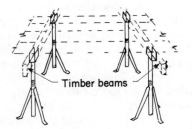

Timber beams

Adjustable steel splitheads

Fig. 11.8   Rigid trestles

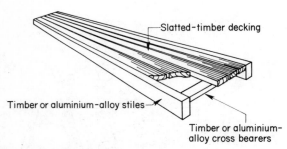

Slatted-timber decking

Timber or aluminium-alloy stiles

Timber or aluminium-alloy cross bearers

Fig. 11.9   Proprietary staging

11.4 Tubular scaffold (steel)

Tubular metal 'components' together with a variety of 'couplers and fittings' can be used to assemble various scaffolding forms. Three of the most common are

i)   putlog scaffolding (Fig. 11.10),
ii)   independent tied scaffolding (Fig. 11.11),
iii) access towers (Fig. 11.12), which may be mounted on castors for mobility.

Each form is designed and constructed to suit individual job requirements. The function of each tubular component is stated in Table 11.2, and Fig. 11.13 locates some of them. Table 11.3 lists and describes the location and/or function of those couplers and fittings which are commonly available – most of which are illustrated in Fig. 11.14.

Note   Steel and aluminium alloy tubes should not be used in the same scaffold structure, because of their different strength qualities.

Putlog scaffold (Fig. 11.10)

This has on its outer side a single row of standards normally set at between 1.8 m to 2.4 m apart (depending on the load) which stand off solid or compacted ground via baseplates and a timber sole-plate. Inner support is provided by the wall of the building into which the blade end of the putlogs (or adaptors) are built-in. The distance between putlogs will depend on whether they support a working platform and the thickness of the scaffold board; for example, under a platform decked with 38 mm thick boards, they should be set about 1.2 m apart (see Table 11.4). Where boards abut, putlogs are doubled (as shown in Fig. 11.16). Putlogs are connected to horizontal ledgers. Ledgers which are fixed to the standards are positioned at vertical intervals of about 1.35 m (for bricklayers) to 2.0 m which would provide for headroom between lifts if under platforms were to be used.

Longitudinal or facade braces at about 45° are required over the whole height of the scaffold to form either a vee or a zig-zag pattern.

All ties should preferably be the 'through' type – similar to that shown in Fig. 11.15(a) or the cast-in anchor type such as the one shown in Fig. 11.15(f) – but, until the walls are stable enough to hold the scaffold, temporary raking struts should be set-up from the ground (Fig. 11.15(b)).

Working platforms (Fig. 11.16) should not be more than five boards wide. An arrangement for a toe board and guard rail, which must be provided where a person is liable to fall more than 2.0 m, is shown in Fig. 11.16.

Independent tied scaffold (Fig. 11.11)

This consists of a double row of standards which stand off solid or compacted ground via baseplates and a timber sole-plate. The inner row may be set not more than 300 mm away from the building. For use as a general-purpose scaffold, the distance between rows should not normally exceed the width of five boards. Pairs of standards may be set between 2.0 m and 2.4 m apart, depending upon loading. Transoms are spaced as for putlogs, and transom ends are connected to ledgers. The first pair of ledgers is positioned not more than 2.0 m above the ground; further pairs at intervals of 2.0 m.

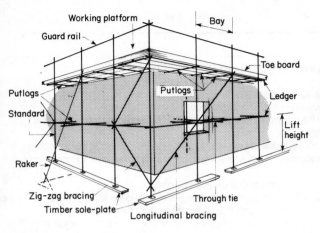

Fig. 11.10 Putlog-scaffold arrangement

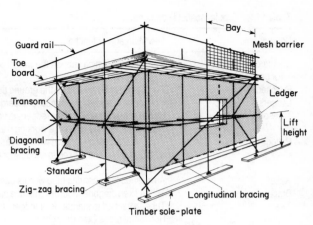

Fig. 11.11 Independent-tied-scaffold arrangement

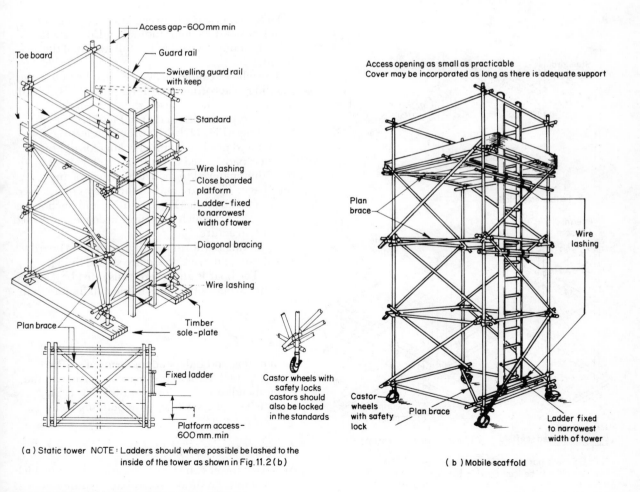

(a) Static tower NOTE: Ladders should where possible be lashed to the
inside of the tower as shown in Fig. 11.2 (b)

(b) Mobile scaffold

Fig. 11.12 Tower scaffold (use of steel tubes and fittings)

Table 11.2　Tubular scaffold components

Components	Definition and/or function
Standards (or uprights)	Vertical tubes which support the weight of the scaffold and its load.
Puncheons	Vertical tubes not based on the ground but starting from within the scaffold.
Ledgers	Horizontal tubes which connect standards longitudinally and may support putlogs and transoms.
Transoms	Horizontal tubes which span ledgers to tie the scaffold transversely. Together with intermediate transoms (board bearers), they may support a working platform.
Putlogs	Horizontal tubes, one end of which is flattened or connected with a putlog end to bear on or be built into a supporting wall. The other end is fixed to a ledger or a standard. Together with intermediate putlogs, they may support a working platform.
Raker	An inclined load-bearing tube with its baseplate bearing against a suitable foundation.
Braces	Tubes fixed diagonally across the face and width and through a scaffold to give stability to the whole scaffold.
Ties	Tubes attached to standards or ledgers (by right-angle couplers) to secure the scaffold to the building.
Bridles	Horizontal tubes fixed (by right-angle couplers) across wall openings to provide support for a putlog transom or tie tube.

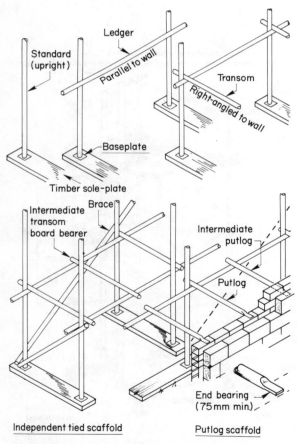

N.B. Couplers not shown (all load bearing couplers to be load bearing fittings)

Fig. 11.13　Location of tubular components

Diagonal bracing between alternate pairs of standards and 45° longitudinal and/or zig-zag bracing should extend to the full height of the scaffold.

The whole scaffold structure must be tied to the building in accordance with BS 5973 : 1981. Ideally at least half of the ties used should be of the through type similar to that in Fig. 11.15(c). Reveal ties (Fig. 11.15(d)) provide a less reliable anchor and need constant checking for tightness. Timber 'pads', used to protect the reveals, should be of thin material to reduce the effect of any shrinkage. There are many other forms of ties used, such as the 'box' tie (Fig. 11.15(e)) and anchor ties (Fig. 11.15(f)) which may have been cast or drilled into the structure.

A working platform together with a toe board, guard rail, and brick guard are shown in Fig. 11.16. Toe boards and guard rails must be provided where a person is liable to fall more than 2.0 m.

Ties

Both putlog and independent forms of scaffold should be tied to the walls in accordance with the recommendations set down in BS 5973 : 1981, Section 9 – tying scaffolding to building facades.

Access towers (Fig. 11.12)

An access tower can be either stationary – in which case it stands on timber sole-plates – or mobile, with castors at each corner.

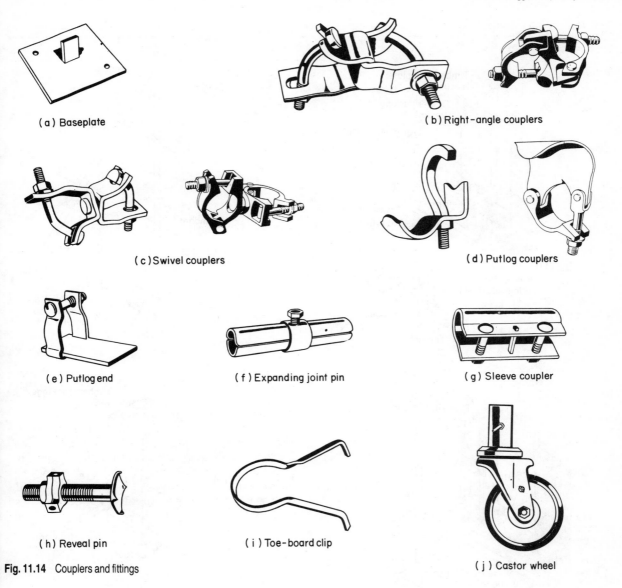

(a) Baseplate

(b) Right-angle couplers

(c) Swivel couplers

(d) Putlog couplers

(e) Putlog end

(f) Expanding joint pin

(g) Sleeve coupler

(h) Reveal pin

(i) Toe-board clip

(j) Castor wheel

Fig. 11.14 Couplers and fittings

The structure is made up of normal steel tubing and couplers. Corner standards should be not less than 1.2 m apart, and each lift should not exceed the width of the shortest side to a maximum lift height of 2.7 m. Diagonal bracing is carried out in zig-zag style at each side, over the full height of the tower. Plan bracing is introduced across the base and alternate lifts.

Height limits will depend on the type (static or mobile) and the situation (exterior or interior use). For example, mobile access towers used externally should not be higher than 3 times the width of the narrowest base side; for interior use, the height could be extended to 3½ times that width. In situations where operatives are to apply horizontal forces, for example drilling operations, provision must be made to prevent the tower overturning.

Toe boards and guard rails should be provided to all sides as shown in Fig. 11.12.

The ground under a mobile scaffold must be level and firm. While static, the castors should be facing outwards from the tower and be locked (brake on). No attempt should be made to move a mobile tower until the platform is totally clear of operatives and their equipment. Movement pressure should then only be applied from the base.

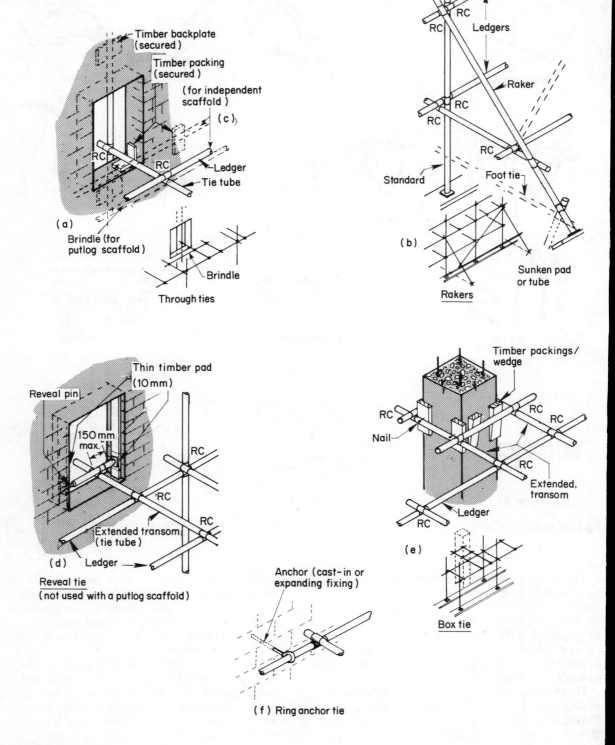

Fig. 11.15 Ties and rakers

Table 11.3 Scaffold couplers and fittings

Couplers and fittings	Fig. 11.14	Definition and/or function
Baseplate	(a)	A square metal plate with a spigot over which a standard or raker is positioned to distribute the load over a greater area.
Adjustable baseplate	—	A baseplate capable of height adjustment via a threaded shaft and collar.
Right-angle coupler (90° coupler)	(b)	For joining scaffold tube at right angles – used for fixing ledgers, transoms, or putlogs to standards, and for securing ties and braces etc.
Swivel coupler	(c)	Used for joining scaffold tubes at any angle where a right-angle coupler cannot be used.
Putlog coupler	(d)	Used for joining intermediate putlogs and transoms to ledgers.
Putlog end	(e)	A fitting to convert a scaffold tube into a putlog.
Joint pin	(f)	Used for joining scaffold tube end-to-end – it fits internally, expanding to grip the tube. Joint pins may only be used for standards or puncheons – although a sleeve coupler is preferred.
Sleeve coupler	(g)	Joins scaffold tubes end-to-end, externally. Used to join ledgers, braces, all tower-scaffold members, and recommended for standards.
Reveal pin	(h)	An adjustable screw fitting used for tightening a tube serving as part of a reveal tie.
Toe-board clip	(i)	For fixing toe boards to scaffold members.
Castor wheels	(j)	Lockable swivelling wheels which are secured to the base of vertical members, e.g. standards, used in conjunction with a mobile scaffold.

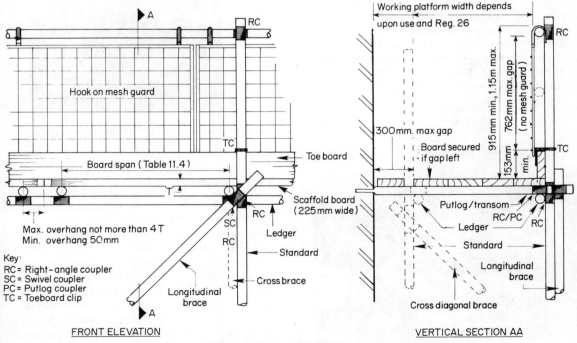

FRONT ELEVATION

VERTICAL SECTION AA

NOTE :– Where possible all bracing should be carried out by the use of R/A couplers. This not always possible though, to keep sensible access

Fig. 11.16 Working platform for a tubular scaffold

Table 11.4   Scaffold-board span in relation to its thickness

Nominal thickness of board (mm)	Maximum span (m)	Minimum overhang (mm)	Maximum overhang (mm)
38	1.5	50	150
50	2.6	50	200
63	3.25	50	250

Note   See Fig. 11.16

11.5 System scaffold

These systems (Figs 11.17 and 11.18), which can be adapted to suit all types of scaffolding situations, are either fully or partly standardised.

A fully standardised system consists of a number of interlocking components, each with its own built-in fixing device (slot, or pin and wedge) to simplify assembly.

Partly standardised systems may, for example, use a series of 'H' frames in place of standards and main transoms, the rest of the scaffold being made up of traditional tubes, couplers, and fittings. 'Easyfix' scaffold systems use fixed-length transoms and putlogs with couplers permanently attached to them – the rest of the scaffold is traditional. Partly standardised systems therefore offer the flexibility of traditional scaffolding plus some of the advantages of a full system.

(a)

(b)

Fig. 11.18   GKN Kwickform 'Quickstage' scaffold system

Tower systems

Figure 11.19 shows an assembled aluminium mobile tower in use – notice that access to the working platform is via a built-in integral stair at each lift. Figure 11.20(a) shows a static tower with outriggers to provide greater stability, thereby allowing the working platform height to be increased (as stated by the manufacturers). Static towers can in some cases be linked together as shown in Figure 11.20(b). Access to the working platform is in this case via a hinged trap door, which forms part of the working platform. All tower access systems must only be climbed from within the inside of the tower, using the built-in stair or ladder. There is a code of practice for the erection of aluminium scaffold produced by the

Fig. 11.17   GKN Kwickform 'Speedy Scaf' frame system scaffold

Prefabricated Aluminium Scaffold Manufacturers Association (PASMA). This is available from all leading scaffold hirers.

General

Erection and dismantling procedures for systems scaffolds should be carried out according to the maker's instructions and statutory legislation.

11.6 Inspection of scaffolding

Scaffolds should be inspected by the constructor before they are handed over for use. Thereafter all types of scaffolding must be inspected at least once every week by a competent person, to ensure that all couplers are tight, braces and ties are secure, and platforms, their toe boards, guard rails, and access ladders are in good condition and held secure. Scaffold foundations should likewise be checked.

Particular attention should be paid to checking the whole structure, or system, after inclement weather – for example after strong winds, changes in temperature, or frost as this may have affected its safety or stability.

All inspections, except those of scaffolds under

Fig. 11.19   GKN Kwickform mobile tower

(b)

Fig. 11.20(a)   Static tower with stabilising outriggers; (b) Two static towers linked together

2 m in height and ladder or trestle scaffolds, must be recorded in a scaffold register (Form F91, part I, Section A). If work lasts less than six weeks the inspection may be telephoned and name of person entered in the register. If lasting more than six weeks then the register is signed by the inspector and kept at the site office. All such reports must be made available to the Health and Safety Inspectorate.

Note   Every employer must ensure that any scaffold used by their employees is safe and complies with the regulations.

1974 No. 903

FACTORIES

The Woodworking Machines Regulations 1974

Made - - - -	*23rd May* 1974
Laid before Parliament	*10th June* 1974
Coming into Operation-- *All Regulations except* *Regulation 41* -	*24th November* 1974
. Regulation 41 - -	*24th May* 1976

ARRANGEMENT OF REGULATIONS

The Secretary of State:—

(*a*) in exercise of powers conferred by sections 17(3), 76 and 180(6) and (7) of the Factories Act 1961(**a**) and now vested in him (**b**) and of all other powers enabling him in that behalf; and

(*b*) after publishing, pursuant to Schedule 4 to the said Act of 1961, notice of the proposal to make the Regulations and after the holding of an inquiry under that Schedule into objections made to the draft,

hereby makes the following Regulations of which all with the exception of Regulation 15 are special Regulations:—

(**a**) 1961 c. 34.   (**b**) S.I. 1968/729 (1968 II, p. 2108).

Part I

Application, Interpretation and Exemptions

Citation, commencement, revocation and amendment

1.—(1) These Regulations may be cited as the Woodworking Machines Regulations 1974 and shall come into operation on 24th November 1974 with the exception of Regulation 41 which shall come into operation on 24th May 1976.

(2) The Regulations specified in columns 1 and 2 of Schedule 2 to these Regulations are hereby revoked to the extent respectively specified in relation thereto in column 3 of that Schedule.

(3) Regulation 67(2) of the Shipbuilding and Ship-repairing Regulations 1960**(a)** and Regulation 42 of the Construction (General Provisions) Regulations 1961**(b)** shall not apply to the parts of woodworking machines required by these Regulations to be guarded or to have other safeguards.

Interpretation

2.—(1) The Interpretation Act 1889**(c)** shall apply to the interpretation of these Regulations as it applies to the interpretation of an Act of Parliament, and as if these Regulations and the Regulations hereby revoked were Acts of Parliament.

(2) In these Regulations, unless the context otherwise requires, the following expressions have the meanings hereby assigned to them respectively, that is to say:—

"approved" means approved for the time being for the purposes of these Regulations by certificate of the Chief Inspector;

"circular sawing machine" means a sawing machine comprising a saw bench (including a bench in the form of a roller table and a bench incorporating a travelling table) with a spindle situated below the machine table to which a circular saw blade can be fitted for the purpose of dividing material into separate parts, but does not include a multiple rip sawing machine, a straight line edging machine or any sawing machine in the operation of which the blade is moved towards the material which is being cut;

"cutters" include saw blades, chain cutters, knives, boring tools, detachable cutters and solid cutters;

"factory" includes any place to which these Regulations apply;

"machine table" includes, in relation to a circular sawing machine, any frame which supports the material being cut;

"narrow band sawing machine" means a sawing machine designed to be fitted with a blade not exceeding 50 millimetres in width in the form of a continuous band or strip the cutting portion of which runs in a vertical direction, but does not include a log band sawing machine or a band re-sawing machine;

"planing machine" means a machine for surfacing or for thicknessing or a combined machine for both those operations but does not include a multi-cutter moulding machine having two or more cutter spindles;

(a) S.I. 1960/1932 (1960 II, p. 1427).   **(b)** S.I. 1961/1580 (1961 II, p. 3207).
(c) 1889 c. 63.

"principal Act" means the Factories Act 1961 as amended by or under any other Act;

"sawmill" means premises which are used solely or mainly for the purpose of sawing logs (including square logs) into planks or boards;

"squared stock" means material having a rectangular (including square) cross section of which the dimensions remain substantially constant throughout the length of the material;

"surfacing" means the planing or smoothing of the surface of material by passing it over cutters and includes chamfering and bevelling, but does not include moulding, tenoning, rebating or recessing;

"vertical spindle moulding machine" includes a high-speed routing machine; and

"woodworking machine" means any machine (including a portable machine) of a kind specified in Schedule 1 to these Regulations for use on all or any one or more of the following, that is to say, wood, cork and fibre board and material composed partly of any of those materials.

Application and operation of Regulations

3.—(1) These Regulations, other than Regulation 15 (which relates to the sale or hire of machinery), shall apply to any of the following, in which any woodworking machine is used, that is to say, to factories and to any premises, places, processes, operations and works to which the provisions of Part IV of the principal Act with respect to special regulations for safety and health are applied by any of the following provisions of that Act, namely, section 123 (which relates to electrical stations), section 124 (which relates to institutions), section 125 (which relates to certain dock premises and certain warehouses), section 126 (which relates to ships) and section 127 (which relates to building operations and works of engineering construction).

(2) In relation to the parts of woodworking machines required by these Regulations to be guarded or to have other safeguards, the provisions of these Regulations as respects guarding and the provision of other safeguards are in substitution for the provisions of section 14(1) of the principal Act and accordingly the provisions of that subsection shall not apply in relation to any such parts.

(3) The provisions of Regulation 12 are in substitution for section 3(1) of the principal Act and accordingly the provisions of that subsection shall not apply in relation to any room to which that Regulation applies.

(4) Except as provided in paragraphs (2) and (3) of this Regulation, the provisions of these Regulations shall be in addition to and not in substitution for the provisions of the principal Act.

Exemptions

4. The Chief Inspector may (subject to such conditions, if any, as may be specified therein) by certificate in writing (which he may in his discretion revoke at any time) exempt from all or any of the requirements of these Regulations—

(a) any particular woodworking machine or any type of woodworking machine; or

(b) any operation or process or any class or description of operations or processes; or

(c) any factory or any part of any factory or any class or description of factories or parts thereof,

if he is satisfied that the requirements in respect of which the exemption is granted are not necessary for the protection of persons employed. Where such exemption is granted, a legible copy of the certificate, showing the conditions (if any) subject to which it has been granted, shall be kept posted in any factory in which the exemption applies in a position where it may be conveniently read by the persons employed.

PART II

ALL WOODWORKING MACHINES—GENERAL

Provision and construction of guards

5.—(1) Without prejudice to the other provisions of these Regulations, the cutters of every woodworking machine shall be enclosed by a guard or guards to the greatest extent that is practicable having regard to the work being done threat, unless the cutters are in such position as to be as safe to every person employed as they would be if so enclosed.

(2) All guards provided in pursuance of the foregoing paragraph of this Regulation shall be of substantial construction.

Adjustment of machines and guards

6. No person shall, while the cutters are in motion—

(a) make any adjustment to any guard on a woodworking machine unless means are provided whereby such an adjustment can be made without danger; or

(b) make any adjustment to any part of a woodworking machine, except where the adjustment can be made without danger.

Use and maintenance of guards, etc.

7.—(1) At all times while the cutters are in motion, the guards and devices required by these Regulations and all such safeguards as are mentioned in Regulation 8 shall be kept constantly in position and properly secured and adjusted except when, and to the extent to which, because of the nature of the work being done, the use of any such guard, device, or safeguard is rendered impracticable:

Provided that the said exception shall not apply to the use of any guard required by Regulations 18(1), 21(1) or (2), 22(1), 23, 28, 30 or 31.

(2) The said guards, devices, and safeguards, and all such appliances as are mentioned in Regulation 14(1)(b) shall be properly maintained.

Exception from obligations to provide guards, etc.

8. Regulations 5, 16, 21, 22, 26, 28, 30, 31 and 36 shall not apply to any machine in respect of which other safeguards are provided which render the machine as safe as it would be if the provisions of those Regulations were complied with.

Machine controls

9. Every woodworking machine shall be provided with an efficient device or efficient devices for starting and stopping the machine and the control or controls of the device or devices shall be in such a position and of such design and construction as to be readily and conveniently operated by the person operating the machine.

Working space

10. There shall be provided around every woodworking machine sufficient clear and unobstructed space to enable, in so far as is thereby practicable, the work being done at the machine to be done without risk of injury to persons employed.

Floors

11. The floor or surface of the ground around every woodworking machine shall be maintained in good and level condition and, as far as reasonably practicable, free from chips and other loose material and shall not be allowed to become slippery.

Temperature

12.—(1) Subject to the following provisions of this Regulation, effective provision shall be made for securing and maintaining a reasonable temperature in every room or other place (not in the open air) in which a woodworking machine is being worked.

(2) In that part of any room or other place (not in the open air) in which a woodworking machine is being worked, a temperature of less than 13 degrees Celsius shall not be deemed at any time to be a reasonable temperature except where and in so far as the necessities of the business carried on make it impracticable to maintain a temperature of at least 13 degrees Celsius.

(3) Where it is impracticable for the aforesaid reasons to maintain a temperature of at least 13 degrees Celsius in any such part of a room or place as aforesaid, there shall be provided in the said part, to the extent that is reasonably practicable, effective means of warming persons working there.

(4) There shall not be used in any such room or place as aforesaid any heating appliance other than an appliance in which the heating element or flame is so enclosed within the body of the appliance that there is no likelihood of the accidental ignition of any material in that room or place by reason of contact with or proximity to the heating element or any flame, except where the heating appliance is so positioned or protected that there is no such likelihood.

(5) Paragraphs (2) and (3) of this Regulation shall in their application to parts of factories which are used as sawmills have effect as if for the references to 13 degrees Celsius there were substituted references to 10 degrees Celsius.

(6) No method of heating shall be employed which results in the escape into the air of any such room or place as aforesaid of any fume of such a character and to such extent as to be likely to be injurious or offensive to persons employed therein.

Training

13.—(1) No person shall be employed on any kind of work at a woodworking machine unless—

(*a*) he has been sufficiently trained at machines of a class to which that machine belongs in the kind of work on which he is to be employed; and

(*b*) he has been sufficiently instructed in accordance with paragraph (2) of this Regulation,

except where he works under the adequate supervision of a person who has a thorough knowledge and experience of the working of the machine and of the matters specified in paragraph (2) of this Regulation.

(2) Every person, while being trained to work at a woodworking machine, shall be fully and carefully instructed as to the dangers arising in connection with such machine, the precautions to be observed, the requirements of these Regulations which apply and, in the case of a person being trained to operate a woodworking machine, the method of using the guards, devices and appliances required by these Regulations.

(3) Without prejudice to the foregoing provisions of this Regulation, a person who has not attained the age of 18 years shall not operate any circular sawing machine, any sawing machine fitted with a circular blade, any planing machine for surfacing which is not mechanically fed, or any vertical spindle moulding machine, unless he has successfully completed an approved course of training in the operation of such a machine. Save that where required to do so as part of such a course of training, he may operate such a machine under the adequate supervision of a person who has a thorough knowledge and experience of the working of the machine and of the matters specified in paragraph (2) of this Regulation.

Duties of persons employed

14.—(1) Every person employed shall, while he is operating a woodworking machine—

(*a*) use and keep in proper adjustment the guards and devices provided in accordance with these Regulations and all such safeguards as are mentioned in Regulation 8; and

(*b*) use the spikes, push-sticks, push-blocks, jigs, holders and back stops provided in accordance with these Regulations,

except (in cases other than those specified in the proviso to Regulation 7(1)) when, because of the nature of the work being done, the use of the said guards, devices or other safeguards, or of the appliances mentioned in sub-paragraph (*b*) of this paragraph, is rendered impracticable.

(2) It shall be the duty of every person, being a person employed by the occupier of a factory and trained in accordance with Regulation 13, who discovers any defect in any woodworking machine in that factory or in any guard, device or appliance provided in accordance with these Regulations or in any such safeguard as is mentioned in Regulation 8 (being a defect which may affect the safe working of a woodworking machine) or who discovers that the floor or surface of the ground around any woodworking machine in that factory is not in good and level condition or is slippery, to report the matter without delay to the occupier, manager or other appropriate person.

Sale or hire of machinery

15. The provisions of section 17(2) of the principal Act (which prohibits the sale or letting on hire of certain machines which do not comply with the requirements of that section) shall extend to any woodworking machine which is for use in a factory and which is not provided with such guards or devices as are necessary, and is not so designed and constructed as, to enable any requirement of the following Regulations to be complied with, that is to say, Regulations 9, 16, 17(3), 21, 22, 24, 25, 26, 27, 28, 30, 31 and 39 in so far as the requirement applies to that woodworking machine.

PART III

CIRCULAR SAWING MACHINES

Guarding of circular sawing machines

16.—(1) That part of the saw blade of every circular sawing machine which is below the machine table shall be guarded to the greatest extent that is practicable.

(2) There shall be provided for every circular sawing machine a riving knife which shall be securely fixed by means of a suitable device situated below the machine table, be behind and in a direct line with the saw blade, have a smooth surface, be strong, rigid and easily adjustable and fulfil the following conditions:—

(a) the edge of the knife nearer the saw blade shall form an arc of a circle having a radius not exceeding the radius of the largest saw blade with which the saw bench is designed to be used;

(b) the knife shall be capable of being so adjusted and shall be kept so adjusted that it is as close as practicable to the saw blade, having regard to the nature of the work being done, and so that at the level of the machine table the distance between the edge of the knife nearer to the saw blade and the teeth of the saw blade does not exceed 12 millimetres;

(c) for a saw blade of a diameter of less than 600 millimetres, the knife shall extend upwards from the machine table to a height above the machine table which is not more than 25 millimetres below the highest point of the saw blade, and for a saw blade of a diameter of 600 millimetres or over, the knife shall extend upwards from the machine table to a height of at least 225 millimetres above the machine table; and

(d) in the case of a parallel plate saw blade the knife shall be thicker than the plate of the saw blade.

(3) Without prejudice to the requirements of Regulation 18(1), that part of the saw blade of every circular sawing machine which is above the machine table shall be guarded with a strong and easily adjustable guard, which shall be capable of being so adjusted and shall be kept so adjusted that it extends from the top of the riving knife to a point above the upper surface of the material being cut which is as close as practicable to that surface or, where squared stock is being fed to the saw blade by hand, to a point which is not more than 12 millimetres above the upper surface of the material being cut.

(4) The guard referred to in the last foregoing paragraph shall have a flange of adequate depth on each side of the saw blade and the said guard shall be kept so adjusted that the said flanges extend beyond the roots of the teeth of the saw blade. Where the guard is fitted with an adjustable front extension piece, that extension piece shall have along the whole of its length a flange of adequate depth on the side remote from the fence and the said extension piece shall be kept so adjusted that the flange extends beyond the roots of the teeth of the saw blade:

Provided that in the case of circular sawing machines manufactured before the date of the coming into operation of this Regulation, the requirements of this paragraph shall not apply until two years after the said date and in the case of such machines, until the expiration of the said period, the said guard shall have along the whole of its length a flange of adequate depth on the side remote from the fence and shall be kept so adjusted that the said flange extends beyond the roots of the teeth of the saw blade.

Sizes of circular saw blades

17.—(1) In the case of a circular sawing machine the spindle of which is not capable of being operated at more than one working speed, no saw blade shall be used thereat for dividing material into separate parts which has a diameter of less than six-tenths of the diameter of the largest saw blade with which the saw bench is designed to be used.

(2) In the case of a circular sawing machine which has arrangements for the spindle to operate at more than one working speed, no saw blade shall be used thereat for dividing material into separate parts which has a diameter of less than six-tenths of the diameter of the largest saw blade which can properly be used at the fastest working speed of the spindle at that saw bench.

(3) There shall be securely affixed to every circular sawing machine a notice specifying the diameter of the smallest saw blade which may be used in the machine in compliance with paragraph (1) or (2) (as the case may be) of this Regulation.

Limitations on the use of circular sawing machines for certain purposes

18.—(1) No circular sawing machine shall be used for cutting any rebate, tenon, mould or groove, unless that part of the saw blade or other cutter which is above the machine table is effectively guarded.

(2) No circular sawing machine shall be used for a ripping operation (other than any such operation involved in cutting a rebate, tenon, mould or groove) unless the teeth of the saw blade project throughout the operation through the upper surface of the material being cut.

(3) No circular sawing machine shall be used for cross-cutting logs or branches unless the material being cut is firmly held by a gripping device securely fixed to a travelling table.

Provision of push-sticks

19.—(1) A suitable push-stick shall be provided and kept available for use at every circular sawing machine which is fed by hand.

(2) Except where the distance between a circular saw blade and its fence is so great or the method of feeding material to the saw blade is such that the use of a push-stick can safely be dispensed with, the push-stick so provided shall be used—

(*a*) to exert feeding pressure on the material between the saw blade and the fence throughout any cut of 300 millimetres or less in length;

(*b*) to exert feeding pressure on the material between the saw blade and the fence during the last 300 millimetres of any cut of more than 300 millimetres in length; and

(*c*) to remove from between the saw blade and the fence pieces of material which have been cut.

Removal of material cut by circular sawing machines

20.—(1) Except as provided in paragraph (3) of this Regulation, where any person (other than the operator) is employed at a circular sawing machine in removing while the saw blade is in motion material which has been cut, that person shall not for that purpose stand elsewhere than at the delivery end of the machine.

(2) Except as provided in paragraph (3) of this Regulation, where any person (other than the operator) is employed at a circular sawing machine in removing while the saw blade is in motion material which has been cut, the machine table shall be constructed or shall be extended over its whole width (by the provision of rollers or otherwise) so that the distance between the delivery end of the table or of any such extension thereof and the up-running part of the saw blade is not less than 1200 millimetres. Provided that this requirement shall not apply to moveable machines which cannot accommodate a blade having a diameter of more than 450 millimetres.

(3) The requirements of paragraphs (1) and (2) of this Regulation shall not apply to a circular sawing machine having a saw bench in the form of a roller table or a saw bench incorporating a travelling table which (in either case) is in motion during the cutting operation.

Part IV

Multiple Rip Sawing Machines and Straight Line Edging Machines

Multiple rip sawing machines and straight line edging machines

21.—(1) Every multiple rip sawing machine and straight line edging machine shall be provided on the operator's side of the in-feed pressure rollers with a suitable device which shall be of such design and so constructed as to contain so far as practicable any material accidentally ejected by the machine and every such device shall extend for not less than the full width of the said pressure rollers.

(2) Every multiple rip sawing machine and straight line edging machine on which the saw spindle is mounted above the machine table shall, in addition to the device required to be provided under paragraph (1) of this Regulation, be fitted on the side remote from the fence with a suitable guard, which shall extend from the edge of the said device along a line parallel to the blade of the saw at least 300 millimetres towards the axis of the saw and shall be of such a design and so constructed as to contain as far as practicable any material accidentally ejected from the machine.

(3) In the case of multiple rip sawing machines and straight line edging machines manufactured before the date of the coming into operation of this Regulation, the requirements of this Regulation shall not apply until two years after the said date.

PART V

NARROW BAND SAWING MACHINES

Narrow band sawing machines

22.—(1) The saw wheels of every narrow band sawing machine and the whole of the blade of every such machine, except that part of the blade which runs downwards between the top wheel and the machine table, shall be enclosed by a guard or guards of substantial construction.

(2) That part of the blade of every such machine as aforesaid which is above the friction disc or rollers and below the top wheel shall be guarded by a frontal plate which is as close as is practicable to the saw blade and has at least one flange at right angles to the plate and extending behind the saw blade.

(3) The friction disc or rollers of every such machine as aforesaid shall be kept so adjusted that they are as close to the surface of the machine table as is practicable having regard to the nature of the work being done.

PART VI

PLANING MACHINES

Limitation on the use of planing machines

23. No planing machine shall be used for cutting any rebate, recess, tenon or mould unless the cutter is effectively guarded.

Cutter blocks for planing machines for surfacing

24. Every planing machine for surfacing which is not mechanically fed shall be fitted with a cylindrical cutter block.

Table gap

25.—(1) Every planing machine for surfacing which is not mechanically fed shall be so designed and constructed as to be capable of adjustment so that the clearance between the cutters and the front edge of the delivery table does not exceed 6 millimetres (measured radially from the centre of the cutter block) and the gap between the feed table and the delivery table is as small as practicable having regard to the operation being performed, and no such planing machine which is not so adjusted shall be used for surfacing.

(2) In the case of planing machines manufactured before the date of the coming into operation of this Regulation, the requirements of the foregoing paragraph of this Regulation shall not apply until twelve months after the said date.

Provision of bridge guards

26.—(1) Every planing machine for surfacing which is not mechanically fed shall be provided with a bridge guard which shall be strong and rigid, have a length not less than the full length of the cutter block and a width not less than the diameter of the cutter block and be so constructed as to be capable of easy adjustment both in a vertical and horizontal direction.

(2) Every bridge guard provided in pursuance of paragraph (1) of this Regulation shall be mounted on the machine in a position which is approximately central over the axis of the cutter block and shall be so constructed as to prevent its being accidentally displaced from that position.

(3) In the case of planing machines manufactured before the date of the coming into operation of this Regulation, the requirements of this Regulation shall not apply until twelve months after the said date, and until the expiration of the said period such machines for surfacing shall be provided with a bridge guard capable of covering the full length and breadth of the cutting slot in the bench and so constructed as to be easily adjusted both in a vertical and horizontal direction.

Adjustment of bridge guards

27.—(1) While a planing machine which is not mechanically fed is being used for surfacing, the bridge guard provided in pursuance of Regulation 26 shall be so adjusted as to enable, so far as is thereby practicable, the work being done at the machine to be done without risk of injury to persons employed.

(2) Except as provided in paragraph (4) of this Regulation and in Regulation 29, when a wider surface of squared stock is being planed or smoothed, the bridge guard so provided shall be adjusted so that the distance between the end of the guard and the fence does not exceed 10 millimetres and the underside of the guard is not more than 10 millimetres above the upper surface of the material.

(3) Except as provided in paragraph (4) of this Regulation, when a narrower surface of squared stock is being planed or smoothed, the bridge guard so provided shall be adjusted so that the end of the guard is at a point not more than 10 millimetres from the surface of the said material which is remote from the fence and the underside of the guard is not more than 10 millimetres above the surface of the feed table.

(4) When the planing or smoothing both of a wider and of a narrower surface of squared stock is being carried out, one operation immediately following the other, the bridge guard so provided shall be adjusted so that when a wider surface is being planed or smoothed the underside of the guard is not more than 10 millimetres above the upper surface of the material and, when a narrower surface is being planed or smoothed, the end of the guard is at a point not more than 10 millimetres from the surface of the said material which is remote from the fence.

(5) Except as provided in paragraph (6) of this Regulation, when the planing of squared stock of square cross section is being carried out, the bridge guard so provided shall be adjusted in a manner which complies with the requirements either of paragraph (2) or of paragraph (3) of this Regulation.

(6) When the planing of two adjoining surfaces of squared stock of square cross section is being carried out, one operation immediately following the other, the bridge guard so provided shall be adjusted so that neither the height of the underside of the guard above the feed table nor the distance between the end of the guard and the fence exceeds the width of the material by more than 10 millimetres.

(7) When the smoothing of squared stock of square cross section is being carried out, the bridge guard so provided shall be adjusted in a manner which complies with the requirements either of paragraph (2) or of paragraph (3) or of paragraph (6) of this Regulation.

Cutter block guards

28.—(1) In addition to being provided with a bridge guard as required by Regulation 26, every planing machine for surfacing which is not mechanically fed shall be provided with a strong, effective and easily adjustable guard for that part of the cutter block which is on the side of the fence remote from the bridge guard.

(2) In the case of planing machines manufactured before the date of the coming into operation of this Regulation, the requirements of the foregoing paragraph of this Regulation shall not apply until twelve months after the said date.

Provision and use of push-blocks

29. When a wider surface of squared stock is being planed or smoothed and by reason of the shortness of the material the work cannot be done with the bridge guard adjusted as required by Regulation 27(2), a suitable push-block having suitable handholds which afford the operator a firm grip shall be provided and used.

Combined machines used for thicknessing

30. That part of the cutter block of a combined machine which is exposed in the table gap shall, when the said machine is used for thicknessing, be effectively guarded.

Protection against ejected material

31.—(1) Every planing machine used for thicknessing shall be provided on the operator's side of the feed roller with sectional feed rollers, or other suitable devices which shall be of such a design and so constructed as to restrain so far as practicable any workpiece ejected by the machine.

(2) Paragraph (1) of this Regulation shall not apply to any machine manufactured before the date of coming into operation of this Regulation; provided that—

    (*a*) not more than one work piece at a time shall be fed to any such machine, and

    (*b*) there shall be securely affixed to every such machine a notice specifying that only single pieces shall be fed.

PART VII

VERTICAL SPINDLE MOULDING MACHINES

Construction, maintenance and mounting of cutters etc.

32. Every detachable cutter for any vertical spindle moulding machine shall be of the correct thickness for the cutter block or spindle on which it is to be mounted and shall be so mounted as to prevent it, so far as practicable, from becoming accidentally detached therefrom.

Provision of false fences

33. Where straight fences are being used for the purposes of the work being done at a vertical spindle moulding machine, the gap between the fences shall be reduced as far as practicable by a false fence or otherwise.

Provision of jigs or holders

34. Where by reason of the nature of the work being done at a vertical spindle moulding machine it is impracticable to provide in pursuance of Regulation 5 a guard enclosing the cutters of the said machine to such an extent that they are effectively guarded, but it is practicable to provide, in addition to the guard required to be provided by Regulation 5, a jig or holder of such a design and so constructed as to hold firmly the material being machined and having suitable handholds which afford the operator a firm grip, the machine shall not be used unless such a jig or holder is provided.

Design and construction of guards for protection against ejected parts

35. Every guard provided in pursuance of Regulation 5 for the cutters of any vertical spindle moulding machine shall be of such a design and so constructed as to contain, so far as reasonably practicable, any part of the cutters or their fixing appliances or any part thereof in the event of their ejection.

Provision and use of back stops

36. Where the work being done at a vertical spindle moulding machine is work in which the cutting of the material being machined commences otherwise than at the end of a surface of the said material and it is impracticable to provide a jig or holder in pursuance of Regulation 34, the trailing end of the said material shall if practicable be supported by a suitable back stop where this would prevent the said material being thrown back when the cutters first make contact with it.

Limitation on the use of vertical spindle moulding machines

37. No work shall be done on a vertical spindle moulding machine being work in which the cutting of the material being machined commences otherwise than at the end of a surface of the said material and during the progress of the cutting the material is moved in the same direction as the movement of the cutters, unless a jig or holder provided in pursuance of Regulation 34 is being used.

Provision of spikes or push-sticks

38. Where the nature of the work being performed at a vertical spindle moulding machine is such that the use of a suitable spike or push-stick would enable the work to be carried on without unnecessary risk, such a spike or push-stick shall be provided and kept available for use.

Machines driven by two speed motors

39.—(1) Where the motor driving a vertical spindle moulding machine (other than a high-speed routing machine) is designed to operate at two working speeds the device controlling the speed of the motor shall be so arranged that the motor cannot run at the higher of those speeds, without first running at the lower of those speeds.

(2) In the case of machines manufactured before the coming into operation of this Regulation, the requirements of the foregoing paragraph of this Regulation shall not apply until twelve months after the said date.

PART VIII

EXTRACTION EQUIPMENT AND MAINTENANCE

Cleaning of saw blades

40. The blade of a sawing machine shall not be cleaned by hand while the blade is in motion.

Extraction of chips and other particles

41. Effective exhaust appliances shall be provided and maintained at every planing machine used for thicknessing other than a combined machine for surfacing and thicknessing, every vertical spindle moulding machine, every multi-cutter moulding machine, every tenoning machine and every automatic lathe, for collecting from a position as close to the cutters as practicable and to the extent that is practicable, the chips and other particles of material removed by the action of the cutters and for discharging them into a suitable receptacle or place:

Provided that this Regulation shall not apply to any high-speed routing machine which incorporates means for blowing away from the cutters the chips or particles as they are removed or to either of the following which is not used for more than six hours in any week, that is to say, any vertical spindle moulding machine and any tenoning machine.

Maintenance and fixing

42.—(1) Every woodworking machine and every part thereof, including cutters and cutter blocks, shall be of good construction, sound material and properly maintained.

(2) Every woodworking machine, other than a machine which is held in the hand, shall be securely fixed to a foundation, floor, or to a substantial part of the structure of the premises, save that where this is impracticable, other arrangements shall be made to ensure its stability.

PART IX

LIGHTING

Lighting

43. In addition to the requirements of subsections (1) and (4) of section 5 of the principal Act and the Factories (Standards of Lighting) Regulations 1941**(a)**, the following provisions shall have effect in respect of any work done with any woodworking machine:—

    (*a*) the lighting, whether natural or artifical, for every woodworking machine shall be sufficient and suitable for the purpose for which the machine is used;

    (*b*) the means of artificial lighting for every woodworking machine shall be so placed or shaded as to prevent glare and so that direct rays of light do not impinge on the eyes of the operator while he is operating such machine.

(a) S. R. & O. 1941/94 (Rev. VII, p. III: 1941 I, p. 280).

PART X

Signed by order of the Secretary of State.

Harold Walker,
Joint Parliamentary Under Secretary of State,
Department of Employment.

23rd May 1974.

SCHEDULE 1

Regulation 2(2)

Machines which are woodworking machines for the purposes of these Regulations

1. Any sawing machine designed to be fitted with one or more circular blades.
2. Grooving machines.
3. Any sawing machine designed to be fitted with a blade in the form of a continuous band or strip.
4. Chain sawing machines.
5. Mortising machines.
6. Planing machines.
7. Vertical spindle moulding machines (including high-speed routing machines).
8. Multi-cutter moulding machines having two or more cutter spindles.
9. Tenoning machines.
10. Trenching machines.
11. Automatic and semi-automatic lathes.
12. Boring machines.

Regulation 1(2)   SCHEDULE 2

Column 1 Regulations revoked	Column 2 References	Column 3 Extent of Revocation
1. The Woodworking Machinery Regulations 1922.	S.R. & O. 1922/1196 (Rev. VII, p. 458: 1922, p. 273).	The whole Regulations.
2. The Woodworking Machinery (Amendment) Regulations 1927.	S.R. & O. 1927/207 (Rev. VII, p. 462: 1927, p. 440).	The whole Regulations.
3. The Woodworking (Amendment of Scope) Special Regulations 1945.	S.R. & O. 1945/1227 (Rev. VII, p. 462: 1945 I, p. 380).	The whole Regulations.
4. The Railway Running Sheds (No. 2) Regulations 1961.	S.I. 1961/1768 (1961 III, p. 3410).	In the Schedule, the items numbered 2, 5 and 9.

Index